Activ Online Physics provides interactive visualizations of processes and representations to help you better understand key physical processes, experiment quantitatively, and develop your critical-thinking skills. This library of online interactive simulations are coupled with thought-provoking questions and activities to guide your understanding of physics.

Website: www.aw-bc.com/knight

Minimum System Requirements:
Windows: 250 MHz; OS 98, NT, ME, 2000, XP
Macintosh: 233 MHz; OS 9.2, 10
Both:
- 64 RAM installed
- 1024 x 768 screen resolution
- Browsers: Internet Explorer 5.5, 6.0; Netscape 6.2.3, 7.0
- Plug Ins: Macromedia's Flash 6.2.3

MasteringPhysics™ is the first Socratic tutoring system developed specifically for physics students like you. It is the result of years of detailed studies of how students work physics problems, and where they get stuck and need help. Studies show students who used MasteringPhysics significantly improved their scores on traditional final exams and the Force Concept Inventory (a conceptual test) when compared with traditional hand-graded homework.

With your purchase of a new copy of Knight's *Physics for Scientists and Engineers,* you should have received a Student Access Kit for **MasteringPhysics™** if your professor required it as a component of your course. The kit contains instructions and a code for you to access MasteringPhysics.

If you did not purchase a new textbook and your professor requires you to enroll in the **MasteringPhysics** online homework and tutorial program, you may purchase an online subscription with a major credit card. Go to www.masteringphysics.com and follow the links to purchasing online.

Minimum System Requirements:
Windows: 250 MHz; OS 98, NT, ME, 2000, XP
Macintosh: 233 MHz; OS 9.2, 10
RedHat Linux 8.0
All:
- 64 RAM installed
- 1024 x 768 screen resolution
- Browsers: Internet Explorer 5.0, 5.5, 6.0; Netscape 6.2.3, 7.0; Mozilla 1.2, 1.3

MasteringPhysics™ is powered by MyCyberTutor by Effective Educational Technologies

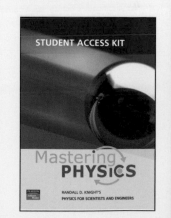

STUDENT ACCESS KIT

Mastering PHYSICS

RANDALL D. KNIGHT'S
PHYSICS FOR SCIENTISTS AND ENGINEERS

AW003

Table of Problem-Solving Strategies

Note for users of the five-volume edition:
Volume 1 (pp. 1–481) includes chapters 1–15.
Volume 2 (pp. 482–607) includes chapters 16–19.
Volume 3 (pp. 608–779) includes chapters 20–24.
Volume 4 (pp. 780–1194) includes chapters 25–36.
Volume 5 (pp. 1148–1383) includes chapters 36–42.

Chapters 37–42 are not in the Standard Edition.

Physics for Scientists and Engineers

A Strategic Approach

ActivPhysics™ OnLine Activities

Activ ONLINE Physics   www.aw-bc.com/knight

Physics for Scientists and Engineers

and Engineers Volume 2

A Strategic Approach

Randall D. Knight

California Polytechnic State University, San Luis Obispo

PEARSON

Addison
Wesley

San Francisco   Boston   New York
Cape Town   Hong Kong   London   Madrid   Mexico City
Montreal   Munich   Paris   Singapore   Sydney   Tokyo   Toronto

Executive Editor:	Adam Black, Ph.D.
Development Editor:	Alice Houston, Ph.D.
Project Manager:	Laura Kenney Editorial & Production Services
Associate Editor:	Liana Allday
Media Producer:	Claire Masson
Marketing Manager:	Christy Lawrence
Market Development:	Susan Winslow
Manufacturing Supervisor:	Vivian McDougal
Art Director:	Blakely Kim
Production Service:	Thompson Steele, Inc.
Text Design:	Mark Ong, Side by Side Studios
Cover Design:	Yvo Riezebos Design
Illustrations:	Precision Graphics
Photo Research:	Cypress Integrated Systems
Cover Printer:	Phoenix Color Corporation
Printer and Binder:	R. R. Donnelley & Sons

Cover Image:	Rainbow/PictureQuest
Credits:	see page C–1

Library of Congress Cataloging-in-Publication Data

Knight, Randall Dewey.
   Physics for scientists and engineers : a strategic approach / Randall D. Knight.
     p. cm.
   Includes index.
   ISBN 0-8053-8960-1 (extended ed. with MasteringPhysics)
    1. Physics   I. Title.

QC23.2.K65 2004
530--dc22

2003062809

ISBN 0-8053-8967-9 Volume 2 with MasteringPhysics
ISBN 0-8053-9009-x Volume 2 without MasteringPhysics

6 7 8 9 10—DOW—07
www.aw-bc.com

Brief Contents

About the Author

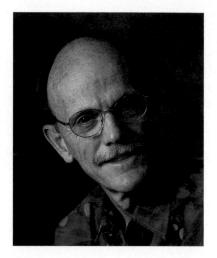

Randy Knight has taught introductory physics for over 20 years at Ohio State University and California Polytechnic University, where he is currently Professor of Physics. Professor Knight received a bachelor's degree in physics from Washington University in St. Louis and a Ph.D. in physics from the University of California, Berkeley. He was a post-doctoral fellow at the Harvard-Smithsonian Center for Astrophysics before joining the faculty at Ohio State University. It was at Ohio State that he began to learn about the research in physics education that, many years later, led to this book.

Professor Knight's research interests are in the field of lasers and spectroscopy, and he has published over 25 research papers. He recently led the effort to establish an environmental studies program at Cal Poly, where, in addition to teaching introductory physics, he also teaches classes on energy, oceanography, and environmental issues. When he's not in the classroom or in front of a computer, you can find Randy hiking, sea kayaking, playing the piano, or spending time with his wife Sally and their seven cats.

Preface to the Instructor

In 1997 we published *Physics: A Contemporary Perspective.* This was the first comprehensive, calculus-based textbook to make extensive use of results from physics education research. The development and testing that led to this book had been partially funded by the National Science Foundation. In the preface we noted that it was a "work in progress" and that we very much wanted to hear from users—both instructors and students—to help us shape the book into a final form.

And hear from you we did! We received feedback and reviews from roughly 150 professors and, especially important, 4500 of their students. This textbook, the newly titled *Physics for Scientists and Engineers: A Strategic Approach*, is the result of synthesizing that feedback and using it to produce a book that we hope is uniquely tuned to helping today's students succeed. It is the first introductory textbook built from the ground up on research into how students can more effectively learn physics.

Objectives

My primary goals in writing *Physics for Scientists and Engineers: A Strategic Approach* have been:

- To produce a textbook that is more focused and coherent, less encyclopedic.
- To move key results from physics education research into the classroom in a way that allows instructors to use a range of teaching styles.
- To provide a balance of quantitative reasoning and conceptual understanding, with special attention to concepts known to cause student difficulties.
- To develop students' problem-solving skills in a systematic manner.
- To support an active-learning environment.

These goals and the rationale behind them are discussed at length in my small paperback book, *Five Easy Lessons: Strategies for Successful Physics Teaching* (Addison Wesley, 2002). Please request a copy from your local Addison Wesley sales representative if it would be of interest to you (ISBN 0-8053-8702-1).

Textbook Organization

The 42-chapter extended edition (ISBN 0-8053-8685-8) of *Physics for Scientists and Engineers* is intended for use in a three-semester course. Most of the 36-chapter standard edition (ISBN 0-8053-8982-2), ending with relativity, can be covered in two semesters, but the judicious omission of a few chapters will avoid rushing through the material and give students more time to develop their knowledge and skills.

There's a growing sentiment that quantum physics is quickly becoming the province of engineers, not just scientists, and that even a two–semester course should include a reasonable introduction to quantum ideas. The *Instructor's Guide* outlines a couple of routes through the book that allow most of the quantum physics chapters to be reached in two semesters. I've written the book with the hope that an increasing number of instructors will choose one of these routes.

- **Extended edition,** with modern physics (ISBN 0-8053-8685-8): chapters 1–42.
- **Standard edition** (ISBN 0-8053-8982-2): chapters 1–36.
- **Volume 1** (ISBN 0-8053-8963-6) covers mechanics: chapters 1–15.
- **Volume 2** (ISBN 0-8053-8966-0) covers thermodynamics: chapters 16–19.
- **Volume 3** (ISBN 0-8053-8969-5) covers waves and optics: chapters 20–24.
- **Volume 4** (ISBN 0-8053-8972-5) covers electricity and magnetism, plus relativity: chapters 25–36.
- **Volume 5** (ISBN 0-8053-8975-X) covers relativity and quantum physics: chapters 36–42.
- **Volumes 1–5** boxed set (ISBN 0-8053-8978-4).

The full textbook is divided into seven parts: Part I: *Newton's Laws*, Part II: *Conservation Laws*, Part III: *Applications of Newtonian Mechanics*, Part IV: *Thermodynamics*, Part V: *Waves and Optics*, Part VI: *Electricity and Magnetism*, and Part VII: *Relativity and Quantum Mechanics*. Although I recommend covering the parts in this order (see below), doing so is by no means essential. Each topic is self-contained, and Parts III–VI can be rearranged to suit an instructor's needs. To facilitate a reordering of topics, the full text is available in the five individual volumes listed in the margin.

Organization Rationale: Thermodynamics is placed before waves because it is a continuation of ideas from mechanics. The key idea in thermodynamics is energy, and moving from mechanics into thermodynamics allows the uninterrupted development of this important idea. Further, waves introduce students to functions of two variables, and the mathematics of waves is more akin to electricity and magnetism than to mechanics. Thus moving from waves to fields to quantum physics provides a gradual transition of ideas and skills.

The purpose of placing optics with waves is to provide a coherent presentation of wave physics, one of the two pillars of classical physics. Optics as it is presented in introductory physics makes no use of the properties of electromagnetic fields. There's little reason other than historical tradition to delay optics until after E&M. The documented difficulties that students have with optics are difficulties with waves, not difficulties with electricity and magnetism. However, the optics chapters are easily deferred until the end of Part VI for instructors who prefer that ordering of topics.

More Effective Problem-Solving Instruction

Careful and systematic instruction is provided on all aspects of problem solving. Some of the features that support this approach are described here, and more details are provided in the *Instructor's Guide*.

- An emphasis on using *multiple representations*—descriptions in words, pictures, graphs, and mathematics—to look at a problem from many perspectives.
- The explicit use of *models*, such as the particle model, the wave model, and the field model, to help students recognize and isolate the essential features of a physical process.
- TACTICS BOXES for the development of particular skills, such as drawing a free-body diagram or using Lenz's law. Tactics Box steps are explicitly illustrated in subsequent worked examples, and these are often the starting point of a full problem-solving strategy.

TACTICS BOX 4.3 **Drawing a free-body diagram**

❶ **Identify all forces acting on the object.** This step was described in Tactics Box 4.2.
❷ **Draw a coordinate system.** Use the axes defined in your pictorial representation. If those axes are tilted, for motion along an incline, then the axes of the free-body diagram should be similarly tilted.
❸ **Represent the object as a dot at the origin of the coordinate axes.** This is the particle model.
❹ **Draw vectors representing each of the identified forces.** This was described in Tactics Box 4.1. Be sure to label each force vector.
❺ **Draw and label the *net force* vector $\vec{F}_{net}$.** Draw this vector beside the diagram, not on the particle. Or, if appropriate, write $\vec{F}_{net} = \vec{0}$. Then check that $\vec{F}_{net}$ points in the same direction as the acceleration vector $\vec{a}$ on your motion diagram.

TACTICS BOX 32.2 **Evaluating line integrals**

❶ If $\vec{B}$ is everywhere perpendicular to a line, the line integral of $\vec{B}$ is

$$\int_i^f \vec{B} \cdot d\vec{s} = 0$$

❷ If $\vec{B}$ is everywhere tangent to a line of length L *and* has the same magnitude B at every point, the line integral of $\vec{B}$ is

$$\int_i^f \vec{B} \cdot d\vec{s} = BL$$

- **PROBLEM-SOLVING STRATEGIES** that help students develop confidence and more proficient problem-solving skills through the use of a consistent four-step approach: **MODEL, VISUALIZE, SOLVE, ASSESS.** Strategies are provided for each broad class of problems, such as dynamics problems or problems involving electromagnetic induction. The (MP) icon directs students to the specially developed *Skill Builder* tutorial problems in MasteringPhysics™ (see page xi), where they can interactively work through each of these strategies online.

- Worked **EXAMPLES** that illustrate good problem-solving practices through the consistent use of the four-step problem-solving approach and, where appropriate, the Tactics Box steps. The worked examples are often very detailed and carefully lead the student step by step through the *reasoning* behind the solution, not just through the numerical calculations. Steps that are often implicit or omitted in other textbooks, because they seem so obvious to experts, are explicitly discussed since research has shown these are often the points where students become confused.

- **NOTE ▶** Paragraphs within worked examples caution against common mistakes and point out useful tips for tackling problems.

- The *Student Workbook* (see page xi), a unique component of this text, bridges the gap between worked examples and end-of-chapter problems. It provides qualitative problems and exercises that focus on developing the skills and conceptual understanding necessary to solve problems with confidence.

- Approximately 3000 original and diverse *end-of-chapter problems* have been carefully crafted to exercise and test the full range of qualitative and quantitative problem-solving skills. *Exercises*, which are keyed to specific sections, allow students to practice basic skills and computations. *Problems* require a better understanding of the material and often draw upon multiple representations of knowledge. *Challenge Problems* are more likely to use calculus, utilize ideas from more than one chapter, and sometimes lead students to explore topics that weren't explicitly covered in the chapter.

(MP) **PROBLEM-SOLVING STRATEGY 5.2** **Dynamics problems**

MODEL Make simplifying assumptions.

VISUALIZE

Pictorial representation. Show important points in the motion with a sketch, establish a coordinate system, define symbols, and identify what the problem is trying to find. This is the process of translating words to symbols.

Physical representation. Use a motion diagram to determine the object's acceleration vector $\vec{a}$. Then identify all forces acting on the object and show them on a free-body diagram.

It's OK to go back and forth between these two steps as you visualize the situation.

SOLVE The mathematical representation is based on Newton's second law

$$\vec{F}_{net} = \sum_i \vec{F}_i = m\vec{a}$$

The vector sum of the forces is found directly from the free-body diagram. Depending on the problem, either

- Solve for the acceleration, then use kinematics to find velocities and positions, or
- Use kinematics to determine the acceleration, then solve for unknown forces.

ASSESS Check that your result has the correct units, is reasonable, and answers the question.

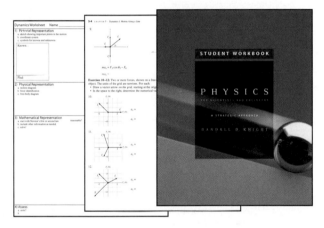

Proven Features to Promote Deeper Understanding

Research has shown that many students taking calculus-based physics arrive with a wealth of misconceptions and subsequently struggle to develop a coherent understanding of the subject. Using a number of unique, reinforcing techniques, this book tackles these issues head-on to enable students to build a solid foundation of understanding.

- A *concrete-to-abstract* approach introduces new concepts through observations about the real world and everyday experience. Step by step, the text then builds up the concepts and principles needed by a theory that will make sense of the observations and make new, testable predictions. This inductive approach better matches how students learn, and it reinforces how physics—and science in general—operates.

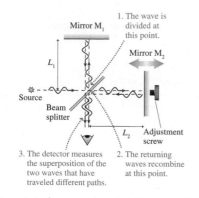

1. The wave is divided at this point.

Mirror M₁

Mirror M₂

L_1

Source

Beam splitter

L_2 Adjustment screw

3. The detector measures the superposition of the two waves that have traveled different paths.

2. The returning waves recombine at this point.

Annotated **FIGURE** showing the operation of the Michelson interferometer.

- **STOP TO THINK** questions embedded in each chapter allow students to assess whether they've understood the main idea of a section. The *Stop to Think* questions, which include concept questions, ratio reasoning, and ranking tasks, are primarily derived from physics education research.

- **NOTE** ▶ paragraphs draw attention to common misconceptions, clarify possible confusions in terminology and notation, and provide important links to previous topics.

- Unique *annotated figures*, based on research into visual learning modes, make the artwork a teaching tool on a par with the written text. Commentary in blue—the "instructor's voice"—helps students "read" the figure. Students "learn by viewing" how to interpret a graph, how to translate between multiple representations, how to grasp a difficult concept through a visual analogy, and many other important skills.

- The learning goals and links that begin each chapter outline what the student needs to remember from previous chapters and what to focus on in the chapter ahead.

  ▶ **Looking Ahead** lists key concepts and skills the student will learn in the coming chapter.

  ◀ **Looking Back** suggests important topics students should review from previous chapters.

- Unique schematic *Chapter Summaries* help students organize their knowledge in an expert-like hierarchy, from general principles (top) to applications (bottom). Side-by-side pictorial, graphical, textual, and mathematical representations are used to help students with different learning styles and enable them to better translate between these key representations.

- *Part Overviews and Summaries* provide a global framework for the student's learning. Each part begins with an overview of the chapters ahead. It then concludes with a broad summary to help students draw connections between the concepts presented in that set of chapters. **KNOWLEDGE STRUCTURE** tables in the part summaries, similar to the chapter summaries, help students see a forest rather than dozens of individual trees.

SUMMARY

The goal of Chapter 27 has been to understand and apply Gauss's law.

GENERAL PRINCIPLES

Gauss's Law

For any *closed* surface enclosing net charge Q_{in}, the net electric flux through the surface is

$$\Phi_e = \oint \vec{E} \cdot d\vec{A} = \frac{Q_{in}}{\epsilon_0}$$

The electric flux Φ_e is the same for *any* closed surface enclosing charge Q_{in}.

Symmetry

The symmetry of the electric field must match the symmetry of the char

In practice, Φ_e is computable only if the symmetry of the Gaussian surface matches the symmetry of the charge distribution.

IMPORTANT CONCEPTS

Charge creates the electric field that is responsible for the electric flux.

Q_{in} is the sum of all enclosed charges. This charge contributes to the flux.

Gaussian surface

Charges outside the surface contribute to the electric field, but they don't contribute to the flux.

Flux is the amount of electric field passing through a surface of area A.

$$\Phi_e = \vec{E} \cdot \vec{A}$$

$\vec{A}$ is the **area vector**.

Surface integrals calculate the flux by summing the fluxes through many small pieces of the surface.

$$\Phi_e = \sum \vec{E} \cdot \delta\vec{A}$$
$$\rightarrow \int \vec{E} \cdot \delta\vec{A}$$

For closed surfaces:
A net flux in or out indicates that the surface encloses a net charge.
Field lines through but with no *net* flux mean that the surface encloses no *net* charge.

Two important situations:
If the electric field is everywhere tangent to the surface, then

$$\Phi_e = 0$$

If the electric field is everywhere perpendicular to the surface *and* has the same strength E at all points, then

$$\Phi_e = EA$$

APPLICATIONS

Conductors in electrostatic equilibrium

- The electric field is zero at all points within the conductor.
- Any excess charge resides entirely on the exterior surface.
- The external electric field is perpendicular to the surface and of magnitude η/ϵ_0, where η is the surface charge density.
- The electric field is zero inside any hole within a conductor unless there is a charge in the hole.

KNOWLEDGE STRUCTURE I   Newton's Laws

ESSENTIAL CONCEPTS	Particle, acceleration, force, interaction
BASIC GOALS	How does a particle respond to a force? How do systems interact?

GENERAL PRINCIPLES	Newton's first law	An object will remain at rest or will continue to move with constant velocity (equilibrium) if and only if $\vec{F}_{net} = \vec{0}$.
	Newton's second law	$\vec{F}_{net} = m\vec{a}$
	Newton's third law	$\vec{F}_{A \text{ on } B} = -\vec{F}_{B \text{ on } A}$

BASIC PROBLEM-SOLVING STRATEGY Use Newton's second law for each particle or system. Use Newton's third law to equate the magnitudes of the two members of an action/reaction pair.

Linear motion	Trajectory motion	Circular motion
$\sum F_x = ma_x$ or $\sum F_x = 0$	$\sum F_x = ma_x$	$\sum F_r = mv^2/r = m\omega^2 r$
$\sum F_y = 0$ $\sum F_y = ma_y$	$\sum F_y = ma_y$	$\sum F_z = 0$

Linear and trajectory kinematics	Circular kinematics
Uniform acceleration: $v_{fs} = v_{is} + a_s \Delta t$ (a_s = constant) $s_f = s_i + v_{is} \Delta t + \frac{1}{2} a_s (\Delta t)^2$ $v_{fs}^2 = v_{is}^2 + 2a_s \Delta s$	**Uniform circular motion:** $\theta_f = \theta_i + \omega \Delta t$ $a_r = v^2/r = \omega^2 r$
Trajectories: The same equations are used for both x and y.	$v = \omega r$
Uniform motion: $s_f = s_i + v_s \Delta t$ ($a = 0$, v_s = constant)	$T = 2\pi r/v = 2\pi/\omega$

General case	$v_s = ds/dt$ = slope of the position graph $a_s = dv_s/dt$ = slope of the velocity graph $v_{fs} = v_{is} + \int_{t_i}^{t_f} a_s \, dt = v_{is}$ + area under the acceleration curve $s_f = s_i + \int_{t_i}^{t_f} v_s \, dt = s_i$ + area under the velocity curve

The Student Workbook

A key component of *Physics for Scientists and Engineers: A Strategic Approach* is the accompanying *Student Workbook*. The workbook bridges the gap between textbook and homework problems by providing students the opportunity to learn and practice skills prior to using those skills in quantitative end-of-chapter problems, much as a musician practices technique separately from performance pieces. The workbook exercises, which are keyed to each section of the textbook, focus on developing specific skills, ranging from identifying forces and drawing free-body diagrams to interpreting wave functions.

The workbook exercises, which are generally qualitative and/or graphical, draw heavily upon the physics education research literature. The exercises deal with issues known to cause student difficulties and employ techniques that have proven to be effective at overcoming those difficulties. The workbook exercises can be used in-class as part of an active-learning teaching strategy, in recitation sections, or as assigned homework. More information about effective use of the *Student Workbook* can be found in the *Instructor's Guide*.

Available versions: Extended (ISBN 0-8053-8961-X), Standard (ISBN 0-8053-8984-9), Volume 1 (ISBN 0-8053-8965-2), Volume 2 (ISBN 0-8053-8968-7), Volume 3 (ISBN 0-8053-8971-7), Volume 4 (ISBN 0-8053-8974-1), and Volume 5 (ISBN 0-8053-8977-6).

Instructor Supplements

- The **Instructor's Guide for Physics for Scientists and Engineers** (ISBN 0-8053-8985-7) offers detailed comments and suggested teaching ideas for every chapter, an extensive review of what has been learned from physics education research, and guidelines for using active-learning techniques in your classroom.

- The **Instructor's Solutions Manuals**, **Chapters 1–19** (ISBN 0-8053-8986-5), and **Chapters 20–42** (ISBN 0-8053-8989-X), written by Professors Pawan Kahol and Donald Foster, at Wichita State University, provide *complete* solutions to all the end-of-chapter problems. The solutions follow the four-step Model/Visualize/Solve/Assess procedure used in the *Problem-Solving Strategies* and all worked examples. Emphasis is placed on the reasoning behind the solution, rather than just the numerical manipulations. The full text of each solution is available as an editable Word document and as a pdf file on the *Instructor's Supplement CD-ROM* for your own use or for posting on your course website.

- The cross-platform **Instructor's Resource CD-ROMs** (ISBN 0-8053-8996-2) consists of the **Simulation and Image Presentation CD-ROM** and the **Instructor's Supplement CD-ROM**. The *Simulation and Image Presentation CD-ROM* provides a comprehensive library of more than 220 applets from *ActivPhysics OnLine*, as well as all the figures from the textbook (excluding photographs) in JPEG format. In addition, all the tables, chapter summaries, and knowledge structures are provided as JPEGs, and the Tactics Boxes, Problem-Solving Strategies, and key (boxed) equations are provided in editable Word format. The *Instructor's Supplement CD-ROM* provides editable Word versions and pdf files of the *Instructor's Guide* and the *Instructor's Solutions Manuals*. Complete *Student Workbook* solutions are also provided as pdf files.

- **MasteringPhysics**™ (www.masteringphysics.com) is a sophisticated, research-proven online tutorial and homework assignment system that provides students with individualized feedback and hints based on their input. It provides a comprehensive library of conceptual tutorials (including one for each

Problem-Solving Strategy in this textbook), multistep self-tutoring problems, and end-of-chapter problems from *Physics for Scientists and Engineers*. *MasteringPhysics*™ provides instructors with a fast and effective way to assign online homework assignments that comprise a range of problem types. The powerful post-assignment diagnostics allow instructors to assess the progress of their class as a whole or to quickly identify individual students' areas of difficulty.

- **ActivPhysics**™ **OnLine** (www.aw-bc.com/knight) provides a comprehensive library of more than 420 tried and tested *ActivPhysics* applets updated for web delivery using the latest online technologies. In addition, it provides a suite of highly regarded applet-based tutorials developed by education pioneers Professors Alan Van Heuvelen and Paul D'Alessandris. The *ActivPhysics* margin icon directs students to specific exercises that complement the textbook discussion.

  The online exercises are designed to encourage students to confront misconceptions, reason qualitatively about physical processes, experiment quantitatively, and learn to think critically. They cover all topics from mechanics to electricity and magnetism and from optics to modern physics. The highly acclaimed *ActivPhysics OnLine* companion workbooks help students work through complex concepts and understand them more clearly. More than 220 applets from the *ActivPhysics OnLine* library are also available on the *Simulation and Image Presentation CD-ROM*.

- The **Printed Test Bank** (ISBN 0-8053-8994-6) and cross-platform **Computerized Test Bank** (ISBN 0-8053-8995-4), prepared by Professor Benjamin Grinstein, at the University of California, San Diego, contain more than 1500 high-quality problems, with a range of multiple-choice, true/false, short-answer, and regular homework-type questions. In the computerized version, more than half of the questions have numerical values that can be randomly assigned for each student.

- The **Transparency Acetates** (ISBN 0-8053-8993-8) provide more than 200 key figures from *Physics for Scientists and Engineers* for classroom presentation.

Student Supplements

- The **Student Solutions Manuals Chapters 1–19** (ISBN 0-8053-8708-0) and **Chapters 20–42** (ISBN 0-8053-8998-9), written by Professors Pawan Kahol and Donald Foster at Wichita State University, provides *detailed* solutions to more than half of the odd-numbered end-of-chapter problems. The solutions follow the four-step Model/Visualize/Solve/Assess procedure used in the *Problem-Solving Strategies* and all worked examples.

- **MasteringPhysics**™ (www.masteringphysics.com) provides students with individualized online tutoring by responding to their wrong answers and providing hints for solving multistep problems. It gives them immediate and up-to-date assessment of their progress, and shows where they need to practice more.

- **ActivPhysics**™ **OnLine** (www.aw-bc.com/knight) provides students with a suite of highly regarded applet-based tutorials (see above). The accompanying workbooks help students work though complex concepts and understand them more clearly. The *ActivPhysics* margin icon directs students to specific exercises that complement the textbook discussion.

- **ActivPhysics OnLine Workbook Volume 1: Mechanics • Thermal Physics • Oscillations & Waves** (ISBN 0-8053-9060-X)

- **ActivPhysics OnLine Workbook Volume 2: Electricity & Magnetism • Optics • Modern Physics** (ISBN 0-8053-9061-8)

■ The **Addison-Wesley Tutor Center** (www.aw.com/tutorcenter) provides one-on-one tutoring via telephone, fax, email, or interactive website during evening hours and on weekends. Qualified college instructors answer questions and provide instruction for *Mastering Physics*™ and for the examples, exercises, and problems in *Physics for Scientists and Engineers*.

Acknowledgments

I have relied upon conversations with and, especially, the written publications of many members of the physics education community. Those who may recognize their influence include Arnold Arons, Uri Ganiel, Ibrahim Halloun, Richard Hake, David Hestenes, Leonard Jossem, Jill Larkin, Priscilla Laws, John Mallinckrodt, Lillian McDermott, Edward "Joe" Redish, Fred Reif, Rachel Scherr, Bruce Sherwood, David Sokoloff, Ronald Thornton, Sheila Tobias, and Alan Van Heuleven. John Rigden, founder and director of the Introductory University Physics Project, provided the impetus that got me started down this path. Early development of the materials was supported by the National Science Foundation as the *Physics for the Year 2000* project; their support is gratefully acknowledged.

I am grateful to Pawan Kahol and Don Foster for the difficult task of writing the *Instructor's Solutions Manuals*; to Jim Andrews and Susan Cable for writing the workbook answers; to Wayne Anderson, Jim Andrews, Dave Ettestad, Stuart Field, Robert Glosser, and Charlie Hibbard for their contributions to the end-of-chapter problems; and to my colleague Matt Moelter for many valuable contributions and suggestions.

I especially want to thank my editor Adam Black, development editor Alice Houston, editorial assistant Liana Allday, and all the other staff at Addison Wesley for their enthusiasm and hard work on this project. Project manager Laura Kenney, Carolyn Field and the team at Thompson Steele, Inc., copy editor Kevin Gleason, photo researcher Brian Donnelly, and page-layout artist Judy Maenle get much of the credit for making this complex project all come together. In addition to the reviewers and classroom testers listed below, who gave invaluable feedback, I am particularly grateful to Wendell Potter and Susan Cable for their close scrutiny of every word and figure.

Finally, I am endlessly grateful to my wife Sally for her love, encouragement, and patience, and to our many cats for their innate abilities to hold down piles of papers and to type qqqqqqqq whenever it was needed.

Randy Knight, September 2003
rknight@calpoly.edu

Reviewers and Classroom Testers

Gary B. Adams, *Arizona State University*
Wayne R. Anderson, *Sacramento City College*
James H. Andrews, *Youngstown State University*
David Balogh, *Fresno City College*
Dewayne Beery, *Buffalo State College*
Joseph Bellina, *Saint Mary's College*
James R. Benbrook, *University of Houston*
David Besson, *University of Kansas*

Randy Bohn, *University of Toledo*
Art Braundmeier, *University of Southern Illinois, Edwardsville*
Carl Bromberg, *Michigan State University*
Douglas Brown, *Cabrillo College*
Ronald Brown, *California Polytechnic State University, San Luis Obispo*
Mike Broyles, *Collin County Community College*

James Carolan, *University of British Columbia*
Michael Crescimanno, *Youngstown State University*
Wei Cui, *Purdue University*
Robert J. Culbertson, *Arizona State University*
Purna C. Das, *Purdue University North Central*
Dwain Desbien, *Estrella Mountain Community College*
John F. Devlin, *University of Michigan, Dearborn*
Alex Dickison, *Seminole Community College*
Chaden Djalali, *University of South Carolina*
Sandra Doty, *Denison University*
Miles J. Dresser, *Washington State University*
Charlotte Elster, *Ohio University*
Robert J. Endorf, *University of Cincinnati*
Tilahun Eneyew, *Embry-Riddle Aeronautical University*
F. Paul Esposito, *University of Cincinnati*
John Evans, *Lee University*
Michael R. Falvo, *University of North Carolina*
Abbas Faridi, *Orange Coast College*
Stuart Field, *Colorado State University*
Daniel Finley, *University of New Mexico*
Jane D. Flood, *Muhlenberg College*
Thomas Furtak, *Colorado School of Mines*
Richard Gass, *University of Cincinnati*
J. David Gavenda, *University of Texas, Austin*
Stuart Gazes, *University of Chicago*
Katherine M. Gietzen, *Southwest Missouri State University*
Robert Glosser, *University of Texas, Dallas*
William Golightly, *University of California, Berkeley*
Paul Gresser, *University of Maryland*
C. Frank Griffin, *University of Akron*
John B. Gruber, *San Jose State University*
Randy Harris, *University of California, Davis*
Stephen Haas, *University of Southern California*
Nicole Herbots, *Arizona State University*
Scott Hildreth, *Chabot College*
David Hobbs, *South Plains College*
Laurent Hodges, *Iowa State University*
John L. Hubisz, *North Carolina State University*
George Igo, *University of California, Los Angeles*
Bob Jacobsen, *University of California, Berkeley*
Rong-Sheng Jin, *Florida Institute of Technology*
Marty Johnston, *University of St. Thomas*
Stanley T. Jones, *University of Alabama*
Darrell Judge, *University of Southern California*
Pawan Kahol, *Wichita State University*
Teruki Kamon, *Texas A&M University*
Richard Karas, *California State University, San Marcos*
Deborah Katz, *U.S. Naval Academy*
Miron Kaufman, *Cleveland State University*
M. Kotlarchyk, *Rochester Institute of Technology*
Cagliyan Kurdak, *University of Michigan*
Fred Krauss, *Delta College*
H. Sarma Lakkaraju, *San Jose State University*

Darrell R. Lamm, *Georgia Institute of Technology*
Robert LaMontagne, *Providence College*
Alessandra Lanzara, *University of California, Berkeley*
Sen-Ben Liao, *Massachusetts Institute of Technology*
Dean Livelybrooks, *University of Oregon*
Chun-Min Lo, *University of South Florida*
Richard McCorkle, *University of Rhode Island*
James McGuire, *Tulane University*
Theresa Moreau, *Amherst College*
Gary Morris, *Rice University*
Michael A. Morrison, *University of Oklahoma*
Richard Mowat, *North Carolina State University*
Taha Mzoughi, *Mississippi State University*
Vaman M. Naik, *University of Michigan, Dearborn*
Craig Ogilvie, *Iowa State University*
Martin Okafor, *Georgia Perimeter College*
Benedict Y. Oh, *University of Wisconsin*
Georgia Papaefthymiou, *Villanova University*
Peggy Perozzo, *Mary Baldwin College*
Brian K. Pickett, *Purdue University, Calumet*
Joe Pifer, *Rutgers University*
Dale Pleticha, *Gordon College*
Robert Pompi, *SUNY-Binghamton*
David Potter, *Austin Community College*
Chandra Prayaga, *University of West Florida*
Didarul Qadir, *Central Michigan University*
Michael Read, *College of the Siskiyous*
Michael Rodman, *Spokane Falls Community College*
Sharon Rosell, *Central Washington University*
Anthony Russo, *Okaloosa-Walton Community College*
Otto F. Sankey, *Arizona State University*
Rachel E. Scherr, *University of Maryland*
Bruce Schumm, *University of California, Santa Cruz*
Douglas Sherman, *San Jose State University*
Elizabeth H. Simmons, *Boston University*
Alan Slavin, *Trent College*
William Smith, *Boise State University*
Paul Sokol, *Pennsylvania State University*
Chris Sorensen, *Kansas State University*
Anna and Ivan Stern, *AW Tutor Center*
Michael Strauss, *University of Oklahoma*
Arthur Viescas, *Pennsylvania State University*
Chris Vuille, *Embry-Riddle Aeronautical University*
Ernst D. Von Meerwall, *University of Akron*
Robert Webb, *Texas A&M University*
Zodiac Webster, *California State University, San Bernardino*
Robert Weidman, *Michigan Technical University*
Jeff Allen Winger, *Mississippi State University*
Ronald Zammit, *California Polytechnic State University, San Luis Obispo*
Darin T. Zimmerman, *Pennsylvania State University, Altoona*

Preface to the Student

From Me to You

The most incomprehensible thing about the universe is that it is comprehensible.
   —Albert Einstein

The day I went into physics class it was death.
   —Sylvia Plath, *The Bell Jar*

Let's have a little chat before we start. A rather one-sided chat, admittedly, because you can't respond, but that's OK. I've heard from many of your fellow students over the years, so I have a pretty good idea of what's on your mind.

What's your reaction to taking physics? Fear and loathing? Uncertainty? Excitement? All of the above? Let's face it, physics has a bit of an image problem on campus. You've probably heard that it's difficult, maybe downright impossible unless you're an Einstein. Things that you've heard, your experiences in other science courses, and many other factors all color your *expectations* about what this course is going to be like.

It's true that there are many new ideas to be learned in physics and that the course, like college courses in general, is going to be much faster paced than science courses you had in high school. I think it's fair to say that it will be an *intense* course. But we can avoid many potential problems and difficulties if we can establish, here at the beginning, what this course is about and what is expected of you—and of me!

Just what is physics, anyway? Physics is a way of thinking about the physical aspects of nature. Physics is not better than art or biology or poetry or religion, which are also ways to think about nature; it's simply different. One of the things this course will emphasize is that physics is a human endeavor. The information content of this book was not found in a cave or conveyed to us by aliens; it was discovered by real people engaged in a struggle with real issues. I hope to convey to you something of the history and the process by which we have come to accept the principles that form the foundation of today's science and engineering.

You might be surprised to hear that physics is not about "facts." Oh, not that facts are unimportant, but physics is far more focused on discovering *relationships* that exist between facts and *patterns* that exist in nature than on learning facts for their own sake. As a consequence, there's not a lot of memorization when you study physics. Some—there are still definitions and equations to learn—but less than in many other courses. Our emphasis, instead, will be on thinking and reasoning. This is important to factor into your expectations for the course.

Perhaps most important of all, *physics is not math!* Physics is much broader. We're going to look for patterns and relationships in nature, develop the logic that relates different ideas, and search for the reasons *why* things happen as they do. In doing so, we're going to stress qualitative reasoning, pictorial and graphical reasoning, and reasoning by analogy. And yes, we will use math, but it's just one tool among many.

It will save you much frustration if you're aware of this physics–math distinction up front. Many of you, I know, want to find a formula and plug numbers into it—that is, to do a math problem. Maybe that's what you learned in high school science courses, but it is *not* what this course expects of you. We'll certainly do

(a) X-ray diffraction pattern

(b) Electron diffraction pattern

many calculations, but the specific numbers are usually the last and least important step in the analysis.

Physics is about recognizing patterns. The top photograph is an x-ray diffraction pattern that shows how a collimated beam of x rays spreads out after passing through a crystal. The bottom photograph shows what happens when a collimated beam of electrons is shot through the same crystal. What does the obvious similarity in these two photographs tell us about the nature of light and about the nature of matter?

As you study, you'll sometimes be baffled, puzzled, and confused. That's perfectly normal and to be expected. Making mistakes is OK too *if* you're willing to learn from the experience. No one is born knowing how to do physics any more than he or she is born knowing how to play the piano or shoot basketballs. The ability to do physics comes from practice, repetition, and struggling with the ideas until you "own" them and can apply them yourself in new situations. There's no way to make learning effortless, at least for anything worth learning, so expect to have some difficult moments ahead.

But also expect to have some moments of excitement at the joy of discovery. There will be instants at which the pieces suddenly click into place and you *know* that you understand a difficult idea. There will be times when you'll surprise yourself by successfully working a difficult problem that you didn't think you could solve. My hope, as an author, is that the excitement and sense of adventure will far outweigh the difficulties and frustrations.

Many of you, I suspect, would like to know the "best" way to study for this course. There is no best way. People are too different, and what works for one student works less effectively for another. But I do want to stress that *reading the text* is vitally important. Class time will be used to clarify difficulties and to develop tools for using the knowledge, but your instructor will *not* use class time simply to repeat information in the text. The basic knowledge for this course is written down within these pages, and the *number one expectation* is that you will read carefully and thoroughly to find and learn that knowledge.

Despite there being no best way to study, I will suggest *one* way that is successful for many students. It consists of the following four steps:

1. **Read each chapter *before* it is discussed in class.** I cannot stress too highly how important this step is. Class attendance is largely ineffective if you have not prepared. When you first read a chapter, focus on learning new vocabulary, definitions, and notation. There's a list of terms and notations at the end of each chapter. Learn them! You won't understand what's being discussed or how the ideas are being used if you don't know what the terms and symbols mean.

2. **Participate actively in class.** Take notes, ask and answer questions, take part in discussion groups. There is ample scientific evidence that *active participation* is far more effective for learning science than is passive listening.

3. **After class, go back for a *careful* rereading of the chapter.** In your second reading, pay closer attention to the details and the worked examples. Look for the *logic* behind each example (and I've tried to help make this clear), not just at what formula is being used. Do the *Student Workbook* exercises for each section as you finish your reading of it.

4. **Finally, apply what you have learned to the homework problems at the end of each chapter.** I strongly encourage you to form a study group with two or three classmates. There's good evidence that students who study regularly with a group do better than the rugged individualists who try to go it alone.

Did someone mention a workbook? The companion *Student Workbook* is a vital part of this course. It contains questions and exercises that ask you to reason *qualitatively*, to use graphical information, and to give explanations. It is through these exercises that you will learn what the concepts mean and will practice the reasoning skills appropriate to the chapter. You will then have acquired the base-line knowledge that you need *before* turning to the end-of-chapter homework problems. In sports or in music, you would never think of performing before you practice, so why would you want to do so in physics? The workbook is where you practice and work on basic skills.

Many of you, I know, would like to go straight to the homework problems and then thumb through the text looking for a formula that seems like it will work. That approach will not succeed in this course, and it's guaranteed to make you frustrated and discouraged. Very few homework problems are "plug and chug" problems where you simply put numbers into a formula. To work the homework problems successfully, you need a better study strategy—either that outlined above or your own—that helps you learn the concepts and the relationships between the ideas. Many of the chapters in this book have Problem-Solving Strategies to help you develop effective problem-solving skills.

A traditional guideline in college is to study two hours outside of class for every hour spent in class, and this text is designed with that expectation. Of course, two hours is an average. Some chapters are fairly straightforward and will go quickly. Others likely will require much more than two study hours per class hour.

Now that you know more about what is expected of you, what can you expect of me? That's a little trickier, because the book is already written! Nonetheless, it was prepared on the basis of what I think my students throughout the years have expected—and wanted—from their physics textbook.

You should know that these course materials—the text and the workbook—are based upon extensive research about how students learn physics and the challenges they face. The effectiveness of many of the exercises has been demonstrated through extensive class testing. I've written the book in an informal style that I hope you will find appealing and that will encourage you to do the reading. And finally, I have endeavored to make clear not only that physics, as a technical body of knowledge, is relevant to your profession but also that physics is an exciting adventure of the human mind.

I hope you'll enjoy the time we're going to spend together.

Detailed Contents

Volume 1 contains chapters 1–15; Volume 2 contains chapters 16-19; Volume 3 contains chapters 20–24; Volume 4 contains chapters 25–36; Volume 5 contains chapters 36–42.

R2-D2's internal systems have been on standby while inside the ice cave on the planet Hoth. How much energy must R2-D2's fuel cells provide to raise the temperature of his internal systems from the standby temperature of −20°C to a working temperature of 30°C? To find out, what properties of R2-D2 do you need to estimate?

Thermodynamics

It's All About Energy

If one area of science is universally applicable, from astrophysics to zoology, it is thermodynamics. Thermodynamics is the science of energy in its broadest context, especially the conversion of energy from one form to another. The word energy itself comes from a Greek root meaning "to do work," and much of thermodynamics is concerned with converting stored energy of various forms, such as the energy in fuels, into the energy called useful work.

Thermodynamics arose hand-in-hand with the industrial revolution as the systematic study of converting heat energy into mechanical motion and work. Hence the name *thermo + dynamics*. Indeed, the analysis of engines and generators of various kinds remains the focus of engineering thermodynamics. But thermodynamics, as a science, now extends to all forms of energy conversions, including those involving living organisms. Here are a few examples:

- Gasoline, diesel, and jet engines convert the energy of a fuel into the mechanical energy of moving pistons, wheels, and gears.
- A generator converts the energy of a fuel into the mechanical energy of a spinning turbine and then into electrical energy.
- A fuel cell converts chemical energy into electrical energy.
- A photovoltaic cell converts the electromagnetic energy of light into electrical energy.
- A laser converts electrical energy into the electromagnetic energy of light.
- An organism converts the chemical energy of food into a variety of other forms of energy, including kinetic energy, sound energy, and thermal energy.

The major goal of Part IV will be to understand how energy transformations such as these take place. Although we can look at only a few applications, these chapters will prepare you to use the ideas of thermodynamics in your own field of study.

Thermodynamics deals with *macroscopic* systems, the prefix *macro* (the opposite of *micro*) meaning large. We will be concerned with systems that are solids, liquids, or gases, rather than the "particles" to which we have grown accustomed. The language of thermodynamics will be of temperature, pressure, volume, and moles rather than of position, velocity, and force. An important task will be to learn this new language.

The questions we will be asking are also different from those that have occupied us until now. Typical questions of thermodynamics are

- How do temperature and pressure change during a certain process?
- How does a physical system such as an engine do mechanical work? How is the work related to the temperature and pressure inside the engine?

483

■ If an engine "burns fuel" to do work, how is the amount of work done related to the fuel energy that was used? Are there limits on how efficient an engine can be?

As you can see, there are many new ideas to learn.

The Micro/Macro Connection

Thermodynamics describes and characterizes a macroscopic system as a whole, regardless of whether the system is a solid crystal, a beaker of liquid, or a container of gas. But as you know, one of the major scientific discoveries of the last two hundred years is that solids, liquids, and gases all consist of atoms. These atoms, each a microscopic particle, are ceaselessly moving about and colliding with each other. Their motion and interactions are described by Newtonian mechanics and, in some cases, quantum physics.

A second important goal of Part IV will be to demonstrate that there is a connection between the *microphysics* of the atoms and molecules in the system and the *macrophysics* of the system as a whole. That is, the macroscopic quantities of thermodynamics can be understood in terms of the microscopic physics of moving atoms and molecules. This link between the realms of the small and the large is called the *micro/macro connection*.

The discovery of this link between the microscopic and the macroscopic is an exciting story and a major success of physics. You will find that the familiar concepts of thermodynamics, such as temperature and pressure, have their roots in atomic-level motion and collisions. You will also find it possible to learn a great deal about the properties of molecules, such as their speeds, on the basis of purely macroscopic measurements. The micro/macro connection will lead to the second law of thermodynamics, one of the most subtle but also one of the most profound and far-reaching statements in physics.

Practical Thermodynamics

Our ultimate destination in Part IV is an understanding of the thermodynamics of *heat engines*. A heat engine is a device, such as a gasoline-powered internal combustion engine, that transforms heat energy into useful work. These are the devices that power our modern society.

Understanding how to transform heat into work will be a significant achievement, but we first have many steps to take along the way. We need to understand the concepts of temperature and pressure. We need to learn about the properties of solids, liquids, and gases. Most important, we need to expand our view of energy to include *heat,* the energy that is transferred between two systems at different temperatures. And at a deeper level, we need to see how these concepts are connected to the underlying microphysics of randomly moving molecules. Only after all these steps have been taken will we be able to analyze a real heat engine.

This is an ambitious goal, but one we can achieve.

16 A Macroscopic Description of Matter

Liquid steel is poured from a giant crucible.

▶ **Looking Ahead**

The goal of Chapter 16 is to learn the characteristics of macroscopic systems. In this chapter you will learn to:

- Understand the basic properties of solids, liquids, and gases.
- Interpret a phase diagram.
- Work with different temperature scales.
- Use the ideal-gas law.
- Understand Ideal-gas processes and represent them on a pV diagram.

◀ **Looking Back**

The material in this chapter depends on thermal energy and the properties of fluids. Please review:

- Section 11.7 Thermal energy.
- Sections 15.1–15.3 Fluids and pressure.

A room full of air, a beaker of water, and a crucible of red-hot molten steel are examples of **macroscopic systems,** systems that are large enough to see or touch. These are the systems of our everyday experience. Our goal in this chapter is twofold:

- To learn what kind of physical properties characterize macroscopic systems.
- To begin the process of connecting a system's macroscopic properties to the underlying motions of the atoms in the system.

The properties of a macroscopic system as a whole are called its **bulk properties.** One fairly obvious example is the system's mass. Other bulk properties to be discussed in this chapter include volume, density, temperature, and pressure. Macroscopic systems are also characterized as being either solid, liquid, or gas. These are called the *phases* of matter, and we'll be interested in when and how a system changes from one phase to another.

Ultimately we would like to understand the macroscopic properties of solids, liquids, and gases in terms of the microscopic motions of their atoms and molecules. Developing this **micro/macro connection** will take several chapters, but we'll start laying the foundations in this chapter. This effort to understand macroscopic properties in terms of the particle-like atoms will pay handsome dividends when we later come to electricity and then quantum physics.

16.1 Solids, Liquids, and Gases

The ice cube you take out of the freezer soon becomes a puddle of liquid water. Then, more slowly, it evaporates to become water vapor in the air. Water is unique. It is the only substance whose three **phases**—solid, liquid, and gas—are familiar from everyday experience.

Each of the elements and most compounds can exist as a solid, liquid, or gas. The change between liquid and solid (freezing or melting) or between liquid and gas (boiling or condensing) is called a **phase change.** Phase changes, which we'll investigate later in this chapter, occur at a well-defined temperature. We're familiar with only one, or perhaps two, of the phases of most substances because their melting point and/or the boiling point are far outside the range of normal human experience. Nonetheless, as the opening photo shows, substances as hard and solid as steel can melt and even boil if the temperature is sufficiently high.

The notion of three distinct phases is less useful for more complex systems. A piece of wood is solid, but liquid wood and gaseous wood don't exist. *Liquid crystals,* which are used to display the numbers on your digital watch, have characteristics of both solids *and* liquids. Complex systems have many interesting properties, but this text will focus on macroscopic systems for which the three phases are distinct.

NOTE ▶ This use of the word *phase* has no relationship at all to the *phase* or *phase constant* of simple harmonic motion and waves. ◀

The three phases of water—solid, liquid, and gas—are familiar from everyday experience.

Solids, liquids, and gases

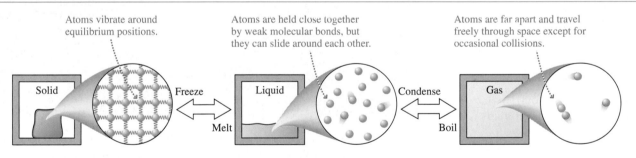

Atoms vibrate around equilibrium positions.

Atoms are held close together by weak molecular bonds, but they can slide around each other.

Atoms are far apart and travel freely through space except for occasional collisions.

Solid — Freeze / Melt — Liquid — Condense / Boil — Gas

A **solid** is a rigid macroscopic system with a definite shape and volume. It consists of particle-like atoms connected together by spring-like molecular bonds. Each atom jiggles and vibrates around an equilibrium position, but an atom is *not* free to move around inside the solid. Solids are nearly *incompressible,* telling us that the atoms in a solid are just about as close together as they can get.

The solid shown here is a **crystal,** meaning that the atoms are arranged in a periodic array. The elements and many compounds have a crystal structure when in their solid phase. In other solids, such as glass, the atoms are frozen into random positions. These are called **amorphous solids.**

A **liquid** is more complicated than either a solid or a gas. Like a solid, a liquid is nearly *incompressible.* This tells us that the molecules in a liquid are about as close together as they can get. Like a gas, a liquid flows and deforms to fit the shape of its container. The fluid nature of a liquid tells us that the molecules are free to move around.

Together, these observations suggest a model in which the molecules of the liquid are loosely held together by weak molecular bonds. The bonds are strong enough that the molecules never get far apart but not strong enough to prevent the molecules from sliding around each other.

A **gas** is a system in which each molecule moves through space as a free, noninteracting particle until, on occasion, it collides with another molecule or with the wall of the container. A gas is a *fluid.* A gas is also highly *compressible,* telling us that there is lots of space between the molecules.

Gases are fairly simple macroscopic systems, hence many of our examples in Part IV will be based on gases.

State Variables

The parameters used to characterize or describe a macroscopic system are known as **state variables** because, taken all together, they describe the *state* of the macroscopic system. You met some state variables in earlier chapters: Volume, pressure, mass, mass density, and thermal energy. We'll soon introduce several new state variables: moles, number density, and, most important, the temperature T.

The state variables are not all independent of each other. For example, you learned in Chapter 15 that a system's mass density ρ is defined in terms of the system's mass M and volume V as

$$\rho = \frac{M}{V} \quad \text{(mass density)} \quad (16.1)$$

In this chapter we'll use an uppercase M for the system mass and a lowercase m for the mass of an atom. Table 16.1 is a short list of mass densities of various substances.

If we change the value of any of the state variables, then we are changing the state of the system. For example, to *compress* a gas means to decrease its volume. Other state variables, such as pressure and temperature, may also change as the volume changes. We will use the symbol Δ to represent a *change* in the value of a state variable. That is, ΔT is a *change* of temperature and Δp is a *change* of pressure. **For any quantity X, ΔX is *always* $X_f - X_i$, the final value minus the initial value.**

A system is said to be in **thermal equilibrium** if its state variables are constant and not changing. As an example, a gas is in thermal equilibrium if it has been left undisturbed long enough for p, V, and T to reach steady values. One of the important goals of Part IV is to establish the conditions under which a macroscopic system reaches thermal equilibrium.

TABLE 16.1 Densities of materials

Substance	ρ (kg/m^3)
Air at STP[*]	1.2
Ethyl alcohol	790
Water (solid)	920
Water (liquid)	1000
Aluminum	2700
Copper	8920
Gold	19,300
Iron	7870
Lead	11,300
Mercury	13,600
Silicon	2330

[*]$T = 0°C$, $p = 1$ atm

EXAMPLE 16.1 The mass of a lead pipe

A project on which you are working uses a cylindrical lead pipe with outer and inner diameters of 4.0 cm and 3.5 cm, respectively, and a length of 50 cm. What is its mass?

SOLVE The mass density of lead is $\rho_{lead} = 11,300$ kg/m^3. The volume of a circular cylinder of length l is $V = \pi r^2 l$. In this case we need to find the volume of the outer cylinder, of radius r_2, *minus* the volume of air in the inner cylinder, of radius r_1. The volume of the pipe is

$$V = \pi r_2^2 l - \pi r_1^2 l = \pi(r_2^2 - r_1^2)l = 1.47 \times 10^{-4} \text{ m}^3$$

Hence the pipe's mass is

$$M = \rho_{lead} V = 1.66 \text{ kg}$$

STOP TO THINK 16.1 The pressure in a system is measured to be 60 kPa. At a later time the pressure is 40 kPa. The value of Δp is

a. 60 kPa.　　　b. 40 kPa.　　　c. 20 kPa.　　　d. -20 kPa.

16.2 Atoms and Moles

The mass of a macroscopic system is directly related to the total number of atoms or molecules in the system, denoted N. Because N is determined simply by counting, it is a number with no units. A typical macroscopic system has $N \sim 10^{25}$ atoms, an incredibly large number.

The symbol $\sim$, if you are not familiar with it, stands for "has the order of magnitude." It means that the number is only known to within a factor of 10 or so. The statement $N \sim 10^{25}$, which is read "N is of order 10^{25}," implies that N is somewhere in the range 10^{24} to 10^{26}. It is far less precise than the "approximately equal" symbol $\approx$. As we begin to deal with large numbers it will often be necessary to distinguish "really large" numbers, such as 10^{25}, from "small" numbers such as a mere 10^5. Saying $N \sim 10^{25}$ gives us a rough idea of how large N is and allows us to know that it differs significantly from 10^5 or even 10^{15}.

It is often useful to know the number of atoms or molecules per cubic meter in a system. We call this quantity the **number density.** It characterizes how densely the atoms are packed together within the system. In an N-atom system that fills volume V, the number density is

$$\frac{N}{V} \quad \text{(number density)} \tag{16.2}$$

The SI units of number density are m^{-3}. The number density of atoms in a solid is $(N/V)_{\text{solid}} \sim 10^{29} \ m^{-3}$. The number density of a gas depends on the pressure, but is usually less than $10^{27} \ m^{-3}$.

The value of N/V in a *uniform* system is independent of the volume V. That is, the number density is the same whether you look at the whole system or just a portion of it. For example, Figure 16.1 shows a $100 \ m^3$ room with 10,000 tennis balls bouncing around. The number density of tennis balls in the room is $N/V = 10,000/100 \ m^3 = 100 \ m^{-3}$. If you were to look at only a one-tenth portion of the room, you would find 1000 balls in $10 \ m^3$, again giving $N/V = 1000/10 \ m^3 = 100 \ m^{-3}$. The number density is the same in both cases.

NOTE ▶ While we might say "There are 100 tennis balls per cubic meter," or "There are 10^{29} atoms per cubic meter," tennis balls and atoms are not units. The units of N/V are simply m^{-3}. ◀

A $100 \ m^3$ room has 10,000 tennis balls bouncing around. The number density of tennis balls in the room is $N/V = 10,000/100 \ m^3 = 100 \ m^{-3}$.

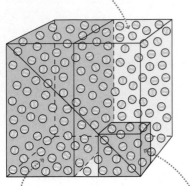

If we look at only half the room, we would find 5000 balls in $50 \ m^3$, again giving $N/V = 5000/50 \ m^3 = 100 \ m^{-3}$.

In one-tenth of the room, we would find 1000 balls in $10 \ m^3$, again giving $N/V = 1000/10 \ m^3 = 100 \ m^{-3}$.

FIGURE 16.1 The number density of a uniform system is independent of the volume.

Atomic Mass and Atomic Mass Number

You will recall from chemistry that atoms of different elements have different masses. The mass of an atom is determined primarily by its most massive constituents, the protons and neutrons in its nucleus. The *sum* of the number of protons and neutrons is called the **atomic mass number** A:

$$A = \text{number of protons} + \text{number of neutrons}$$

A, which by definition is an integer, is written as a leading superscript on the atomic symbol. For example, the common isotope of hydrogen, with one proton and no neutrons, is ^{1}H. The "heavy hydrogen" isotope called *deuterium,* which includes one neutron, is ^{2}H. The primary isotope of carbon, with six protons (which makes it carbon) and six neutrons, is ^{12}C. The radioactive isotope ^{14}C, used for carbon dating of archeological finds, contains six protons and eight neutrons.

The **atomic mass** scale is established by defining the mass of ^{12}C to be exactly 12 u, where u is the symbol for the **atomic mass unit.** That is, $m(^{12}\text{C}) = 12$ u. The atomic mass of any other atom is its mass relative to ^{12}C. For example, careful experiments with hydrogen find that the mass *ratio* $m(^1\text{H})/m(^{12}\text{C})$ is $1.0078/12$. Thus the atomic mass of hydrogen is $m(^1\text{H}) = 1.0078$ u.

The numerical value of the atomic mass of ^{1}H is close to, but not exactly, its atomic mass number $A = 1$. The slight difference is due to the electron mass and to various relativistic effects. For our purposes, it will be sufficient to overlook the slight difference and use the integer atomic mass numbers as the values of the atomic mass. That is, we'll use $m(^1\text{H}) = 1$ u, $m(^4\text{He}) = 4$ u, and $m(^{16}\text{O}) = 16$ u. For molecules, the **molecular mass** is the sum of the atomic masses of the atoms forming the molecule. Thus the molecular mass of the diatomic molecule O_2, the constituent of oxygen gas, is $m(O_2) = 32$ u.

NOTE ▶ An element's atomic mass number is *not* the same as its atomic number. The *atomic number,* which is the element's position in the periodic table, is the number of protons. ◀

Table 16.2 shows the atomic mass numbers of some of the elements that we'll use for examples and homework problems. A complete periodic table, including atomic masses, is found in Appendix B.

Moles and Molar Mass

One way to specify the amount of substance in a macroscopic system is to give its mass. Another way, one connected to the number of atoms, is to measure the amount of substance in *moles*. By definition, one **mole** of matter, be it solid, liquid, or gas, is the amount of substance containing as many basic particles as there are atoms in 12 g of ^{12}C. Many decades of ingenious experiments have determined that there are 6.02×10^{23} atoms in 12 g of ^{12}C, so we can say that 1 mole of substance, abbreviated 1 mol, is 6.02×10^{23} basic particles.

The basic particle depends on the substance. Helium is a **monatomic gas,** meaning that the basic particle is the helium atom. Thus 6.02×10^{23} helium atoms are 1 mol of helium. But oxygen gas is a **diatomic gas** because the basic particle is the two-atom diatomic molecule O_2. 1 mol of oxygen gas contains 6.02×10^{23} *molecules* of O_2 and thus $2 \times 6.02 \times 10^{23}$ oxygen atoms. Table 16.3 lists the monatomic and diatomic gases that we will use for examples and homework problems.

The number of basic particles per mole of substance is called **Avogadro's number,** N_A. The value of Avogadro's number is

$$N_A = 6.02 \times 10^{23} \text{ mol}^{-1}$$

Avogadro's number, like the gravitational constant G, is one of the basic constants of nature.

Despite its name, Avogadro's number is not simply "a number"; it has units. Because there are N_A particles per mole, the number of moles in a substance containing N basic particles is

$$n = \frac{N}{N_A} \tag{16.3}$$

where n is the symbol for moles.

Avogadro's number allows us to determine atomic masses in kilograms. Knowing that N_A ^{12}C atoms have a mass of 12 g, we know the mass of one ^{12}C atom must be

$$m(^{12}\text{C}) = \frac{12 \text{ g}}{6.02 \times 10^{23}} = 1.993 \times 10^{-23} \text{ g} = 1.993 \times 10^{-26} \text{ kg}$$

We defined the atomic mass scale such that $m(^{12}\text{C}) = 12$ u. Thus the conversion factor between atomic mass units and kilograms is

$$1 \text{ u} = \frac{m(^{12}\text{C})}{12} = 1.661 \times 10^{-27} \text{ kg}$$

This conversion factor allows us to calculate the mass in kg of any atom. For example, a ^{20}Ne atom has atomic mass $m(^{20}\text{Ne}) = 20$ u. Multiplying by 1.661×10^{-27} kg/u gives $m(^{20}\text{Ne}) = 3.32 \times 10^{-26}$ kg. If the atomic mass is specified in kilograms, the number of atoms in a system of mass M can be found from

$$N = \frac{M}{m} \tag{16.4}$$

TABLE 16.2 Some atomic mass numbers

Element		A
^{1}H	Hydrogen	1
^{4}He	Helium	4
^{12}C	Carbon	12
^{14}N	Nitrogen	14
^{16}O	Oxygen	16
^{20}Ne	Neon	20
^{27}Al	Aluminum	27
^{40}Ar	Argon	40
^{207}Pb	Lead	207

One mole of helium, sulfur, copper, and mercury.

TABLE 16.3 Monatomic and diatomic gases

Monatomic		Diatomic	
He	Helium	H_2	Hydrogen
Ne	Neon	N_2	Nitrogen
Ar	Argon	O_2	Oxygen

The **molar mass** of a substance is the mass *in grams* of 1 mol of substance. The molar mass, which we'll designate M_{mol}, has units g/mol. By definition, the molar mass of ^{12}C is 12 g/mol. For other substances, whose atomic or molecular masses are given relative to ^{12}C, the numerical value of the molar mass equals the numerical value of the atomic or molecular mass. For example, the molar mass of He, with $m = 4$ u, is $M_{mol}(He) = 4$ g/mol and the molar mass of diatomic O_2 is $M_{mol}(O_2) = 32$ g/mol.

Equation 16.4 uses the atomic mass to find the number of atoms in a system. Similarly, you can use the molar mass to determine the number of moles. For a system of mass M consisting of atoms or molecules with molar mass M_{mol},

$$n = \frac{M \text{ (in grams)}}{M_{mol}} \tag{16.5}$$

NOTE ▶ Equation 16.5 is the one of the few instances where the proper units are *grams* rather than kilograms. ◀

EXAMPLE 16.2 Moles of oxygen

100 g of oxygen gas is how many moles of oxygen?

SOLVE We can do the calculation two ways. First, let's determine the number of molecules in 100 g of oxygen. The diatomic oxygen molecule O_2 has molecular mass $m = 32$ u. Converting this to kg, we get the mass of one molecule:

$$m = 32 \text{ u} \times \frac{1.661 \times 10^{-27} \text{ kg}}{1 \text{ u}} = 5.31 \times 10^{-26} \text{ kg}$$

Thus the number of molecules in 100 g = 0.10 kg is

$$N = \frac{M}{m} = \frac{0.100 \text{ kg}}{5.31 \times 10^{-26} \text{ kg}} = 1.88 \times 10^{24}$$

Knowing the number of molecules gives us the number of moles:

$$n = \frac{N}{N_A} = 3.13 \text{ mol}$$

Alternatively, we can use Equation 16.5 to find

$$n = \frac{M \text{(in grams)}}{M_{mol}} = \frac{100 \text{ g}}{32 \text{ g/mol}} = 3.13 \text{ mol}$$

STOP TO THINK 16.2 Which system contains more atoms: 5 mol of helium ($A = 4$) or 1 mol of neon ($A = 20$)?

a. Helium.   b. Neon.   c. They have the same number of atoms.

16.3 Temperature

We are all familiar with the idea of temperature. You hear the word used nearly every day. But just what is temperature a measure of? Mass is a measure of the amount of substance in a system. Velocity is a measure of how fast a system moves. What physical property of the system have you determined if you measure its temperature?

We will begin with the commonsense idea that temperature is a measure of how "hot" or "cold" a system is. These are properties that we can judge without needing an elaborate theory. As we develop these ideas, we'll find that **temperature** T is related to a system's *thermal energy*. We defined thermal energy in Chapter 10 as the kinetic and potential energy of the atoms and molecules in a system as they vibrate (a solid) or move around (a gas). A system has more thermal energy when it is "hot" than when it is "cold." We'll study temperature more carefully in Chapter 18 and replace these vague notions of hot and cold with a precise relationship between temperature and thermal energy.

To start, we need a means to measure the temperature of a system. This is what a *thermometer* does. A thermometer can be any small macroscopic system that undergoes a measurable change as it exchanges thermal energy with its surroundings. It is placed in contact with a larger system whose temperature it will measure. In a common glass-tube thermometer, for example, a small volume of mercury or alcohol expands or contracts when placed in contact with a "hot" or "cold" object. The object's temperature is determined by the length of the column of liquid.

Other thermometers include:

- Bimetallic strips (two strips of different metals sandwiched together) that curl and uncurl as the temperature changes. These are used in thermostats, such as the one in your house.
- Thermocouples that generate a small voltage depending on the temperature. Thermocouples are widely used for sensing temperatures in inhospitable environments, such as in your car's engine.
- Ideal gases, whose pressure varies with the temperature. We will look at an example of a gas thermometer in a minute.

A thermometer needs a *temperature scale* to be a useful measuring device. In 1742, the Swedish astronomer Anders Celsius sealed mercury into a small capillary tube and observed how it moved up and down the tube as the temperature changed. He selected two temperatures that anyone could reproduce, the freezing and boiling points of pure water, and labeled them 0 and 100. He then marked off the glass tube into one hundred equal intervals between these two reference points. By doing so, he invented the temperature scale that we today call the *Celsius scale*. The units of the Celsius temperature scale are "degrees Celsius," which we abbreviate °C. Note that the degree symbol ° is part of the unit, not part of the number.

NOTE ▶ Because of the 100 equal intervals, the Celsius scale is also called the *centigrade scale.* ◀

The *Fahrenheit scale*, still widely used in the United States, is related to the Celsius scale by

$$T_F = \frac{9}{5}T_C + 32° \tag{16.6}$$

Table 16.4 shows several temperatures measured on the Celsius and Fahrenheit scales and also on the Kelvin scale.

Thermal expansion of the liquid in the thermometer tube pushes it higher in the hot water than in the ice water.

TABLE 16.4 Temperatures measured with different scales

Temperature	T (°C)	T (K)	T (°F)
Melting point of iron	1538	1811	2800
Boiling point of water	100	373	212
Normal body temperature	37.0	310	98.6
Room temperature	20	293	68
Freezing point of water	0	273	32
Boiling point of nitrogen	−196	77	−321
Absolute zero	−273	0	−460

Absolute Zero and Absolute Temperature

(a)

Pressure gauge reading absolute pressure

Rigid gas-filled sphere

System whose temperature is to be measured

T

(b)

Each gas thermometer is calibrated at 0°C and 100°C.

Gas 1

Gas 2

Gas 3

T (°C)

Condensation points

-300 -200 -100 0 100

$T_0 = -273°C$

FIGURE 16.2 The pressure in a constant-volume gas thermometer extrapolates to zero at $T_0 = -273°C$. This is the basis for the concept of absolute zero.

Any physical property that changes with temperature can be used as a thermometer. In practice, the most useful thermometers have a physical property that changes *linearly* with temperature. One of the most important scientific thermometers is the **constant-volume gas thermometer** shown in Figure 16.2a. This thermometer depends on the fact that the *absolute* pressure (not the gauge pressure) of a gas in a sealed container increases linearly as the temperature increases.

A gas thermometer is first calibrated by recording the pressure at two reference temperatures, such as the boiling and freezing points of water. These two points are plotted on a pressure-versus-temperature graph and a straight line is drawn through them. The gas bulb is then brought into contact with the system whose temperature is to be measured. The pressure is measured, then the corresponding temperature is read off the graph.

Figure 16.2b shows the pressure-temperature relationship for three different gases. Notice two important things about this graph.

1. There is a *linear* relationship between temperature and pressure.
2. All gases extrapolate to *zero pressure* at the same temperature: $T_0 = -273°C$. No gas actually gets that cold without condensing, although helium comes very close, but it is surprising that you get the same zero-pressure temperature for any gas and any starting pressure.

The pressure in a gas is due to collisions of the molecules with each other and the walls of the container. A pressure of zero would mean that all motion, and thus all collisions, had ceased. If there were no atomic motion, the system's thermal energy would be zero. The temperature at which all motion would cease, and at which $E_{th} = 0$, is called **absolute zero.** Because temperature is related to thermal energy, absolute zero is the lowest temperature that has physical meaning. We see from the gas-thermometer data that $T_0 = -273°C$. We'll give a somewhat more precise definition of absolute zero in the next section.

It is useful to have a temperature scale with the zero point at absolute zero. Such a temperature scale is called an **absolute temperature scale.** Any system whose temperature is measured on an absolute scale will have $T > 0$. The absolute temperature scale having the same unit size as the Celsius scale is called the *Kelvin scale.* It is the SI scale of temperature. The units of the Kelvin scale are *kelvins,* abbreviated as K. The conversion between the Celsius scale and the Kelvin scale is

$$T_K = T_C + 273 \tag{16.7}$$

NOTE ▶ The units are simply "kelvins," *not* "degrees Kelvin." ◀

On the Kelvin scale, absolute zero is 0 K, the freezing point of water is 273 K, and the boiling point of water is 373 K. While most practical macroscopic devices utilize temperatures in the range ≈ 100 K to ≈ 1000 K, it is worth noting that scientists study the properties of matter from temperatures as low as $\approx 10^{-9}$ K (1 nK) on the one extreme to as high as $\approx 10^{7}$ K on the other!

STOP TO THINK 16.3 The temperature of a glass of water increases from 20°C to 30°C. What is ΔT?

a. 10 K
b. 283 K
c. 293 K
d. 303 K

16.4 Phase Changes

The temperature inside the freezer compartment of a refrigerator is typically about −20°C. Suppose you were to remove a few ice cubes from the freezer and place them in a sealed container with a thermometer. Then, as Figure 16.3a shows, put a steady flame under the container. We'll assume that the heating is done so slowly that the inside of the container always has a single, well-defined temperature.

Figure 16.3b shows the temperature as a function of time. After steadily rising from the initial −20°C, the temperature remains fixed at 0°C for an extended period of time. This is the interval of time during which the ice melts. As it's melting, the ice temperature is 0°C and the liquid water temperature is 0°C. Even though the system is being heated, the liquid water temperature doesn't begin to rise until all the ice has melted. If you were to turn off the flame at any point, the system would remain a mixture of ice and liquid water at 0°C.

> **NOTE** ▶ In everyday language, the three phases of water are called *ice, water,* and *steam.* That is, the term *water* implies the liquid phase. Scientifically, these are the solid, liquid, and gas phases of the compound called *water.* To be clear, we'll use the term *water* in the scientific sense of a collection of H_2O molecules. We'll say either *liquid* or *liquid water* to denote the liquid phase. ◀

The thermal energy of a solid is the kinetic energy of the vibrating atoms plus the potential energy of the stretched and compressed molecular bonds. Melting occurs when the thermal energy gets so large that the molecular bonds begin to break, allowing the atoms to move around. The temperature at which a solid becomes a liquid or, if the thermal energy is reduced, a liquid becomes a solid is called the **melting point** or the **freezing point.** Melting and freezing are *phase changes.*

A system at the melting point is in **phase equilibrium,** meaning that any amount of solid can coexist with any amount of liquid. Raise the temperature ever so slightly and the entire system becomes liquid. Lower it slightly and it all becomes solid. But exactly at the melting point the system has no tendency to move one way or the other. That is why the temperature remains constant at the melting point until the phase change is complete.

You can see the same thing happening in Figure 16.3b at 100°C, the boiling point. This is a phase equilibrium between the liquid phase and the gas phase, and any amount of liquid can coexist with any amount of gas at this temperature. Above this temperature, the thermal energy is too large for bonds to be established between molecules, so the system is a gas. If the thermal energy is reduced, the molecules begin to bond with each other and stick together. In other words, the gas condenses into a liquid. The temperature at which a gas becomes a liquid or, if the thermal energy is increased, a liquid becomes a gas is called the **condensation point** or the **boiling point.**

> **NOTE** ▶ Liquid water becomes solid ice at 0°C, but that doesn't mean the temperature of ice is always 0°C. Ice reaches the temperature of its surroundings. If the air temperature in a freezer is −20°C, then the ice temperature is −20°C. Likewise, steam can be heated to temperatures above 100°C. That doesn't happen when you boil water on the stove because the steam escapes, but steam can be heated far above 100°C in a sealed container. ◀

One difficulty with the experiment of Figure 16.3 is that the outcome depends on the pressure. If you carried out the experiment in Denver, rather than Miami, you would find that the melting point is slightly above 0°C and the boiling is several degrees below 100°C. In other words, the phase-change temperatures are pressure dependent.

(a)

(b)

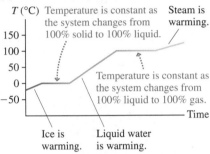

FIGURE 16.3 The temperature as a function of time as water is transformed from solid to liquid to gas.

FIGURE 16.4 Phase diagrams (not to scale) for water and carbon dioxide.

A **phase diagram** is used to show how the phases and phase changes of a substance vary with both temperature and pressure. Figure 16.4 shows the phase diagrams for water and carbon dioxide. You can see that the diagram is divided into three regions corresponding to the solid, liquid, and gas phases. The boundary lines separating the regions indicate the phase transitions. The system is in phase equilibrium at a pressure-temperature point that falls on one of these lines.

Phase diagrams contain a great deal of information. Notice on the water phase diagram that the dotted line at $p = 1$ atm crosses the solid-liquid boundary at 0°C and the liquid-gas boundary at 100°C. These well-known melting and boiling point temperatures of water apply only at standard atmospheric pressure. You can see that in Denver, where $p_{atmos} < 1$ atm, water melts at slightly above 0°C and boils at a temperature below 100°C. A *pressure cooker* works by allowing the pressure inside to exceed 1 atm. This raises the boiling point, so foods that are in boiling water are at a temperature >100°C and cook faster.

In general, crossing the solid-liquid boundary corresponds to melting or freezing while crossing the liquid-gas boundary corresponds to boiling or condensing. But you can see that there's another possibility—crossing the solid-gas boundary. The phase change in which a solid becomes a gas is called **sublimation.** It is not an everyday experience with water because water sublimates only at pressures far below atmospheric pressure. (It does happen in a vacuum chamber, which is how food products are *freeze dried.*) But you are familiar with the sublimation of dry ice. Dry ice is solid carbon dioxide. You can see on the carbon dioxide phase diagram that the dotted line at $p = 1$ atm crosses the solid-*gas* boundary, rather than the solid-liquid boundary, at $T = -78$°C. This is the *sublimation temperature* of dry ice.

Liquid carbon dioxide does exist, but only at pressures greater than 5 atm and temperatures greater than −56°C. A CO_2 fire extinguisher contains *liquid* carbon dioxide under high pressure. (You can hear the liquid slosh if you shake a CO_2 fire extinguisher.)

One important difference between the water and carbon dioxide phase diagrams is the slope of the solid-liquid boundary. For most substances, the solid phase is denser than the liquid phase and the liquid is denser than the gas. Pressurizing the substance compresses it and increases the density. If you start compressing CO_2 gas at room temperature, thus moving upward through the phase diagram along a vertical line, you'll first condense it to a liquid and eventually, if you keep compressing, change it into a solid.

Water is a very unusual substance in that the density of ice is *less* than the density of liquid water. That is why ice floats. If you compress ice, making it denser, you eventually cause a phase transition in which the ice turns to liquid water! Consequently, the solid-liquid boundary for water slopes to the left.

The liquid-gas boundary ends at a point called the **critical point.** Below the critical point, liquid and gas are clearly distinct and there is a phase change if you go from one to the other. But there is no clear distinction between liquid and gas at pressures or temperatures above the critical point. The system is a *fluid,* but it can be varied continuously between high density and low density without a phase change.

The final point of interest on the phase diagram is the **triple point** where the phase boundaries meet. Two phases are in phase equilibrium along the boundaries. The triple point is the *one* value of temperature and pressure for which all three phases can coexist in phase equilibrium. That is, any amounts of solid, liquid, and gas can happily coexist at the triple point. For water, the triple point occurs at $T_3 = 0.01$°C and $p_3 = 0.006$ atm.

The significance of the triple point of water is its connection to the Kelvin temperature scale. The Celsius scale required two *reference points,* the boiling and melting points of water. We can now see that these are not very satisfactory reference points because their values vary as the pressure changes. The Kelvin scale requires only a single reference point. The low end, zero temperature at zero pressure, is fixed by the physical properties of gases. We need only one reference point to determine the slope of the pressure-temperature line of a gas thermometer. And it needs to be a reference that can always be reproduced *with no variation.*

No matter where you are or what the local conditions might be, there's only one temperature at which ice, liquid water, and water vapor will coexist in equilibrium. If you produce this equilibrium in the laboratory, then you *know* the system is at the triple-point temperature. The triple-point temperature of water is an ideal reference point, hence the Kelvin temperature scale is *defined* to be a linear temperature scale starting from 0 K at absolute zero and passing through 273.16 K at the triple point of water. Because $T_3 = 0.01°C$, absolute zero on the Celsius scale is $T_0 = -273.15°C$.

NOTE ▶ To be consistent with our use of significant figures, $T_0 = -273$ K is the appropriate value to use in calculations *unless* you know other temperatures with an accuracy of better than 1°C. ◀

STOP TO THINK 16.4 For which is there a sublimation temperature that is higher than a melting temperature?

a. Water   b. Carbon dioxide   c. Both   d. Neither

16.5 Ideal Gases

Gases are the simplest macroscopic systems. Our goal for the rest of this chapter will be to understand how the macroscopic properties of a gas change as the state of the gas changes.

While we today take it for granted that matter consists of atoms, the evidence for atoms is by no means obvious. The concept of atoms was formulated by two Greek philosophers, Leucippus and Democritus, who flourished about 440–420 BCE. They suggested that all matter consists of small, hard, indivisible, and indestructible particles they called *atoms*. Atoms, they thought, are always in motion, and the spaces between atoms are matter-free regions called the *void*.

Aristotle and most other "mainstream thinkers" were not atomists because they could not accept the idea of the void—or what we, today, would call a vacuum. Consequently, the idea of atoms essentially disappeared for nearly 2000 years. Only with the beginnings of modern chemistry, during the Renaissance, did atomic thinking reappear. Newton was an atomist, but he held the view that atoms in a gas are static, rather than in motion, and that they expand and contract to stay in contact with other atoms. Gases in motion, like the wind, would then be a collective motion of all the atoms moving together with the same velocity.

Daniel Bernoulli advanced a quite different model of gases in 1738. He suggested that a gas consists of small, hard atoms moving randomly at fairly high velocities and, on occasion, colliding with each other or the walls of the container. In other words, Bernoulli reintroduced Democritus' idea of atoms—but updated to include the scientific discoveries of Galileo and Newton! Surprisingly, Bernoulli's ideas were not accepted for nearly a century. The value of his postulates was not recognized until a complete understanding of energy conservation was achieved in the mid-19th century. Numerous scientists then developed Bernoulli's ideas into the kinetic theory of gases that we will study in Chapter 18.

What can macroscopic observations suggest to us about the properties of atoms? One observation, which we noted earlier in the chapter, is that solids and liquids are nearly incompressible. From this we can infer that atoms are fairly "hard" and cannot be pressed together once they come into contact with each other. Atoms also resist being pulled apart. Solids would not be solid if the atoms were not held together by attractive forces. These attractive forces are responsible for the *tensile strength* of solids—how hard you have to pull to break the solid—as well as for the cohesion of liquid droplets. Nonetheless, it is far easier to break a solid or disperse a liquid than it is to compress it, so these attractive forces must be weak in comparison to the repulsive forces that occur when we push the atoms too close together.

These observations imply that an atom is a small particle that is weakly attracted to other nearby atoms but strongly repelled by them if it gets too close. This is precisely the view of molecular bonds that we developed in Chapter 10. Figure 16.5 shows the potential-energy diagram of two atoms separated by distance r. Recall, from Chapter 11, that the force exerted by one atom on the other is the negative of the slope of this graph. The slope is large and negative for values of r less than the equilibrium value r_{eq}, so the force for $r < r_{eq}$ is large and repulsive. For r just slightly greater than r_{eq}, the modest positive slope indicates a weak attractive force. The slope has become nearly zero by $r \approx 0.4$ nm, hence the attractive force is restricted to atoms within about 0.4 nm of each other. Atoms separated by more than about 0.4 nm do not interact.

Solids and liquids are systems in which the atomic separation is very close to r_{eq}, thus the attractive and repulsive atomic forces are balanced. If you try to press the atoms closer together, the repulsive forces resist. If you try to pull them apart, the attractive forces resist.

A gas, by contrast, is much less dense and the average spacing of atoms is much larger than r_{eq}. Consequently, the atoms are usually *not interacting* with each other at all. Instead, they spend most of their time moving freely through space, only occasionally coming close enough to another atom to interact with it. When two atoms collide, it is the steep "wall" of the potential energy curve for $r < r_{eq}$ that is important. That wall represents the repulsive electrical force pushing the atoms apart as they collide. The small distance over which the atoms experience a weak attractive force is of essentially no importance because the atoms spend so little time at those distances.

The Ideal-Gas Model

With these ideas in mind, suppose we were to replace the actual potential-energy curve of Figure 16.5 with the approximate potential-energy curve of Figure 16.6. This is the potential-energy curve for the interaction of two "hard spheres" that have *no* interaction at all until they come into actual contact, at separation $r_{contact}$, and then bounce.

The *hard-sphere model* of the atom represents what we could call the *ideal atom*. It is Democritus' idea of a small, hard particle. A gas of such ideal atoms is called an **ideal gas.** It is a collection of small, hard, randomly moving atoms that occasionally collide and bounce off each other but otherwise do not interact. The ideal gas is a *model* of a real gas and, as with any other model, a simplified description. Nonetheless, experiments show that the ideal-gas model is quite good for gases if two conditions are met:

1. The density is low (i.e., the atoms occupy a volume much smaller than that of the container), and
2. The temperature is well above the condensation point.

If the density gets too high, or the temperature too low, then the attractive forces between the atoms begin to play an important role and our model, which ignores those attractive forces, fails. These are the forces that are responsible, under the right conditions, for the gas condensing into a liquid.

We've been using the term *atoms,* but many gases, as you know, consist of molecules rather than atoms. Only helium, neon, argon, and the other inert elements in the far-right column of the periodic table of the elements form monatomic gases. Hydrogen (H_2), nitrogen (N_2), and oxygen (O_2) are diatomic gases. As far as

FIGURE 16.5 The potential-energy diagram for the interaction of two atoms.

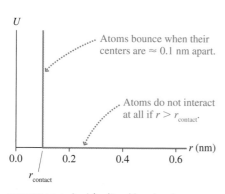

FIGURE 16.6 An idealized hard-sphere model of the interaction potential energy of two atoms.

translational motion is concerned, the ideal-gas model does not distinguish between a monatomic gas and a diatomic gas; both are considered as simply small, hard spheres. Hence the terms *atoms* and *molecules* can be used interchangeably to mean the basic constituents of the gas.

Molecular Speeds

What kind of experimental evidence could distinguish between Newton's view of a gas of static atoms and Bernoulli's model of small atoms in random motion? Suppose we have a container of gas with a very small hole in the side, allowing some of the gas to flow out. If Newton is correct, the atoms should move out uniformly, all touching and moving with the *same speed*—like toothpaste out of a tube as you squeeze it! If Bernoulli is right, the individual atoms should shoot through the hole and emerge with many *different* speeds.

Figure 16.7 shows an experimental apparatus that measures the speeds of molecules in a gas. The molecules emerging from the source form what is called a *molecular beam*. At the right end, a detector counts the number of molecules making it through the apparatus each second. The entire experiment takes place inside a vacuum chamber so that the molecules can travel the distance from source to detector without undergoing any collisions.

The two rotating disks between the source and the detector form a *velocity selector*. Once every revolution, the slot in the first disk allows a small pulse of molecules to pass through. By the time these molecules reach the second disk, the slots have rotated. The molecules can pass through the slot in the second disk and be detected *only* if they have exactly the right speed to travel the distance L between the two disks in exactly the time interval Δt it takes the axle to complete one full revolution. These molecules, with $v = L/\Delta t$, can pass through the second slot as it reappears one revolution later. Molecules having any other speed are blocked by the second disk and are not detected. By changing the rotation frequency of the axle, and thus changing Δt, this apparatus can measure how many molecules have each of many possible speeds.

Figure 16.8 shows the results of an experiment performed with nitrogen gas (N_2) at $T = 20°C$. The results are presented in the form of a histogram. A **histogram** is a bar chart in which the height of each bar tells how many (or, in this case, what percent) of the molecules have a speed in the *range* of speeds shown below the bar. For example, 16% of the molecules have speeds in the range from 600 m/s to 700 m/s. All the bars added together sum to 100%, showing that this histogram describes *all* of the molecules leaving the hole.

Newton's view, in which all the molecules would have the same speed, is clearly incorrect. Instead, we see what is called a *distribution* of speeds, ranging from as low as ≈ 0 m/s to as high as ≈ 1100 m/s. This is exactly as expected for a Bernoulli gas of individual molecules moving randomly at many different speeds. Thus Figure 16.8 is convincing evidence that the basic assumptions behind the ideal-gas model are correct.

But we can learn more. You can see from Figure 16.8 that not all speeds are equally likely. There is a most probable speed of ≈ 550 m/s. This is really fast, ≈ 1200 mph! But also notice that the majority of molecular speeds do not differ much from the most probable speed. There are a few with very high or very low speeds, but well over 60% (sum of the center four bars) of the molecules have speeds within the range 300 m/s to 700 m/s.

The gas in the source consists of a very large number of molecules, each moving randomly and with a different speed. Yet the *collective* behavior of this sample of atoms is exceedingly stable and predictable. If we repeated this experiment over and over, we would get the results of Figure 16.8 every time. This idea—that the *average* properties of a very large number of molecules is a predictable quantity—will underlie our approach to the micro/macro connection in Chapter 18.

The only molecules that reach the detector are those whose speed allows them to travel distance L during the time it takes the disks to make one full revolution.

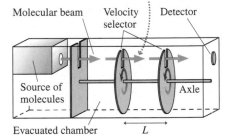

FIGURE 16.7 An apparatus to measure the speeds of molecules in a gas.

FIGURE 16.8 A histogram showing the distribution of speeds in a beam of N_2 molecules at $T = 20°C$.

The Ideal-Gas Law

Section 16.1 introduced the idea of *state variables,* those parameters that describe the state of a macroscopic system. The state variables for an ideal gas are the volume V of its container, the number of moles n of the gas present in the container, the temperature T of the gas and its container, and the pressure p that the gas exerts on the walls of the container. These four state parameters are not independent of each other. If you change the value of one—by, say, raising the temperature—then one or more of the others will change as well. Each change of the parameters is a *change of state* of the system.

Experiments during the 17th and 18th centuries showed that there is a very specific relationship between the four state variables. Suppose you measure p, V, n, and T of a gas, with the temperature measured in kelvins. Then change the state of the gas, by heating it or compressing it or doing something else to it, and measure p, V, n, and T again. Repeat this many times, changing the state of the gas each time, until you have a large table of p, V, n, and T values.

Then make a graph on which you plot pV, the product of the pressure and volume, on the vertical axis and nT, the product of the number of moles and temperature, on the horizontal axis. The very surprising result is that for *any* gas, whether it is hydrogen or helium or oxygen or methane, **you get exactly the same graph,** the linear graph shown in Figure 16.9. In other words, nothing about the graph indicates what gas was used because all gases give the same result.

NOTE ▶ No real gas could extend to $nT = 0$ because it would condense. But an ideal gas never condenses because the only interactions among the molecules are hard-sphere collisions. ◀

As you can see, there is a very clear proportionality between the quantity pV and the quantity nT. If we designate the slope of the line in this graph as R, then we can write the relationship as

$$pV = R \times (nT)$$

It is customary to write this relationship in a slightly different form, namely

$$pV = nRT \qquad \text{(ideal-gas law)} \tag{16.8}$$

Equation 16.8 is the **ideal-gas law.**

The constant R, which is determined experimentally as the slope of the graph in Figure 16.9, is called the **universal gas constant.** Its value, in SI units, is

$$R = 8.31 \text{ J/mol K}$$

The units of R seem puzzling. The denominator mol K is clear because R multiplies nT. But what about the joules? The left side of the ideal-gas law, pV, has units

$$\text{Pa m}^3 = \frac{\text{N}}{\text{m}^2} \text{ m}^3 = \text{N m} = \text{joules}$$

The product pV has units of joules, as shown on the vertical axis in Figure 16.9.

NOTE ▶ You perhaps learned in chemistry to work gas problems using units of atmospheres and liters. To do so, you had a different numerical value of R that was expressed in those units. In physics, however, we always work gas problem in SI units. Pressures *must* be in Pa, volumes in m^3, and temperatures in K before you compute. Calculations using other units will give wildly incorrect answers. ◀

The surprising fact, and one worth commenting upon, is that *all* gases have the *same* graph and the *same* value of R. There is no obvious reason why a very simple

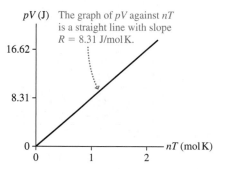

pV (J)   The graph of pV against nT is a straight line with slope $R = 8.31$ J/mol K.

FIGURE 16.9 A graph of pV versus nT for an ideal gas.

atomic gas such as helium should have the same slope as a more complex gas such as methane (CH_4). Nonetheless, both turn out to have the same value for R. The ideal-gas law, within its limits of validity, describes *all* gases with a single value of the constant R.

EXAMPLE 16.3 Calculating a gas pressure

100 g of oxygen gas is distilled into an evacuated 600 cm³ container. What is the gas pressure at a temperature of 150°C?

MODEL The gas can be treated as an ideal gas. Oxygen is a diatomic gas of O_2 molecules.

SOLVE From the ideal-gas law, the pressure is $p = nRT/V$. In Example 16.2 we calculated the number of moles in 100 g of O_2 and found $n = 3.13$ mol. Gas problems typically involve several conversions to get quantities into the proper units, and this example is no exception. The SI units of V and T are m³ and K, respectively, thus

$$V = (600 \text{ cm}^3)\left(\frac{1 \text{ m}}{100 \text{ cm}}\right)^3 = 6.00 \times 10^{-4} \text{ m}^3$$

$$T = (150 + 273) \text{ K} = 423 \text{ K}$$

With this information, the pressure is

$$p = \frac{nRT}{V} = \frac{(3.13 \text{ mol})(8.31 \text{ J/mol K})(423 \text{ K})}{6.00 \times 10^{-4} \text{ m}^3}$$

$$= 1.83 \times 10^7 \text{ Pa} = 181 \text{ atm}$$

In this text we will consider only gases in sealed containers. The number of moles (and number of molecules) will not change during a problem. In that case,

$$\frac{pV}{T} = nR = \text{constant} \tag{16.9}$$

If the gas is initially in state i, characterized by the state variables p_i, V_i, and T_i, and at some later time in a final state f, the state variables for these two states are related by

$$\frac{p_f V_f}{T_f} = \frac{p_i V_i}{T_i} \qquad \text{(ideal gas in a sealed container)} \tag{16.10}$$

This before-and-after relationship between the two states, reminiscent of a conservation law, will be valuable for many problems.

EXAMPLE 16.4 Calculating a gas temperature

A cylinder of gas is at 0°C. A piston compresses the gas to half its original volume and three times its original pressure. What is the final gas temperature?

MODEL Treat the gas as an ideal gas in a sealed container.

SOLVE The before-and-after relationship of Equation 16.10 can be written

$$T_2 = T_1 \frac{p_2}{p_1} \frac{V_2}{V_1}$$

In this problem, the compression of the gas results in $V_2/V_1 = \frac{1}{2}$ and $p_2/p_1 = 3$. The initial temperature is $T_1 = 0°C = 273$ K. With this information,

$$T_2 = 273 \text{ K} \times 3 \times \frac{1}{2} = 409 \text{ K} = 136°C$$

ASSESS We did not need to know actual values of the pressure and volume, just the *ratios* by which they change.

We will often want to refer to the number of molecules N in a gas rather than the number of moles n. This is an easy change to make. Because $n = N/N_A$, the ideal-gas law in terms of N is

$$pV = nRT = \frac{N}{N_A}RT = N\frac{R}{N_A}T \tag{16.11}$$

R/N_A, the ratio of two known constants, is known as **Boltzmann's constant** k_B:

$$k_B = \frac{R}{N_A} = 1.38 \times 10^{-23} \text{ J/K}$$

The subscript B distinguishes Boltzmann's constant from a spring constant or other uses of the symbol k.

Ludwig Boltzmann was an Austrian physicist who did some of the pioneering work in statistical physics during the mid-19th century. Boltzmann's constant k_B can be thought of as the "gas constant per molecule," whereas R is the "gas constant per mole." With this definition, the ideal-gas law in terms of N is

$$pV = Nk_BT \qquad \text{(ideal-gas law)} \tag{16.12}$$

Equations 16.8 and 16.12 are both the ideal-gas law, just expressed in terms of different state variables.

Recall that the number density (molecules per m³) was defined as N/V. A rearrangement of Equation 16.12 gives the number density as

$$\frac{N}{V} = \frac{p}{k_BT} \tag{16.13}$$

This is a useful consequence of the ideal-gas law, but keep in mind that the pressure *must* be in SI units of pascals and the temperature *must* be in SI units of kelvins.

EXAMPLE 16.5 **The distance between molecules**

"Standard temperature and pressure," abbreviated **STP,** are $T = 0°C$ and $p = 1$ atm. Estimate the average distance between gas molecules at STP.

SOLVE Imagine freezing all the molecules at some instant of time. After doing so, place an imaginary cube around each molecule to separate it from all its neighbors. This divides the total volume V of the gas into N small cubes of volume v_i such that the sum of all these small volumes v_i equals the full volume V. Although each of these volumes is somewhat different, we can define an *average* little volume:

$$v_{avg} = \frac{V}{N} = \frac{1}{N/V}$$

That is, the average volume per molecule (m³ per atom) is the inverse of the number of molecules per m³. Note that this is not the volume of the molecule itself, which is much smaller, but the average surrounding volume of space that each molecule can claim as its own, separating it from the other molecules. If we now use Equation 16.13, the number density is

$$\frac{N}{V} = \frac{p}{k_BT} = \frac{1.01 \times 10^5 \text{ Pa}}{(1.38 \times 10^{-23} \text{ J/K})(273 \text{ K})}$$

$$= 2.69 \times 10^{25} \text{ molecules/m}^3$$

where we have used the definition of STP in SI units. Thus the average volume per molecule is

$$v_{avg} = \frac{1}{N/V} = 3.72 \times 10^{-26} \text{ m}^3$$

The volume of a cube is $V = l^3$, where l is the length of each edge. Hence the average length of one of our little cubes is

$$l = (v_{avg})^{1/3} = 3.34 \times 10^{-9} \text{ m} = 3.34 \text{ nm}$$

Because each molecule sits at the center of a cube, the average distance between two molecules is the distance between opposite corners of the cube. As Figure 16.10 shows, this distance is

$$\text{average distance} = \sqrt{l^2 + l^2 + l^2} = \sqrt{3}\,l = 5.7 \text{ nm}$$

The average distance between molecules in a gas at STP is ≈ 5.7 nm.

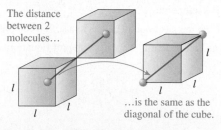

FIGURE 16.10 The distance between two molecules.

The results of this example are important. One of the basic assumptions of the ideal-gas model is that the atoms are "far apart" in comparison to the distance over which atoms exert attractive forces on each other. That distance, as was seen in Figure 16.6, is about 0.4 nm. A gas at STP has an average distance between atoms roughly 14 times the interaction distance. We can safely conclude that the ideal-gas model works well for gases under "typical" circumstances.

STOP TO THINK 16.5 You have two containers of equal volume. One is full of helium gas. The other holds an equal mass of nitrogen gas. Both gases have the same pressure. How does the temperature of the helium compare to the temperature of the nitrogen?

a. $T_{\text{helium}} > T_{\text{nitrogen}}$ b. $T_{\text{helium}} = T_{\text{nitrogen}}$ c. $T_{\text{helium}} < T_{\text{nitrogen}}$

16.6 Ideal-Gas Processes

The ideal-gas law is the connection between the state variables pressure, temperature, and volume. If the state variables change, as they would from heating or compressing the gas, the state of the gas changes. An **ideal-gas process** is the means by which the gas changes from one state to another.

The pV Diagram

It will be very useful to represent ideal-gas processes on a graph called a pV **diagram**. This is nothing more than a graph of pressure versus volume. The important idea behind the pV diagram is that *each point* on the graph represents a single, unique state of the gas. That seems surprising at first, because a point on the graph only directly specifies the values of p and V. But knowing p and V, and assuming that n is known for a sealed container, we can find the temperature by using the ideal-gas law. Thus each point actually represents a triplet of values (p, V, T) specifying the state of the gas.

For example, Figure 16.11a is a pV diagram showing three states of a system consisting of 1 mol of gas. The values of p and V can be read from the axes for each point, then the temperature at that point determined from the ideal-gas law. An ideal-gas process can be represented as a "trajectory" in the pV diagram. The trajectory shows all the intermediate states through which the gas passes. Figures 16.11b and 16.11c show two different processes by which the gas of Figure 16.11a can be changed from state 1 to state 3.

There are infinitely many ways to change the gas from state 1 to state 3. Although the initial and final state are the same for each of them, the particular process by which the gas changes—that is, the particular trajectory—will turn out to have very real consequences. For example, you will soon learn that the work done by an expanding gas, a quantity of very practical importance in various devices, depends on the trajectory followed. The pV diagram is an important graphical representation of the process.

Quasi-Static Processes

Strictly speaking, the ideal-gas law applies only to gases in *thermal equilibrium*. We'll give a careful definition of thermal equilibrium later; for now it is sufficient to say that a system is in thermal equilibrium if its state variables are constant and not changing. Consider an ideal-gas process that changes a gas from state 1 to state 2. The initial and final states are states of thermal equilibrium, with steady values of p, V, and T. But the process, by definition, causes some of these state variables to change. The gas is *not* in thermal equilibrium while the process of changing from state 1 to state 2 is underway.

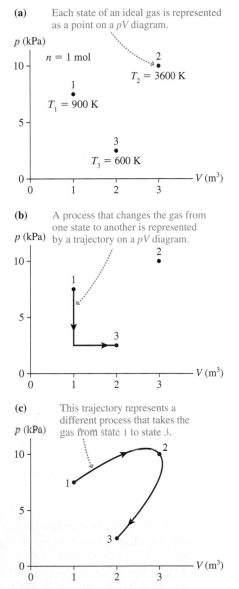

FIGURE 16.11 The state of the gas and ideal-gas processes can be shown on a pV diagram.

(a)

Piston moves
very slowly.

Quasi-static process

(b)

Membrane
suddenly bursts.

Irreversible process

FIGURE 16.12 The slow motion of the piston is a quasi-static process. The bursting of the membrane is not.

To use the ideal-gas law throughout, we will assume that the process occurs *so slowly* that the system is never far from equilibrium. In other words, the values of p, V, and T at any point in the process are essentially the same as the equilibrium values they would assume if we stopped the process at that point. A process that is essentially in thermal equilibrium at all times is called a **quasi-static process.** It is an idealization, like a frictionless surface, but one that is a very good approximation in many real situations.

An important characteristic of a quasi-static process is that the trajectory through the pV diagram can be *reversed*. If you quasi-statically expand a gas by slowly pulling a piston out, as shown in Figure 16.12a, you can reverse the process by slowing pushing the piston in. The gas retraces its pV trajectory until it has returned to its initial state. Contrast this with what happens when the membrane bursts in Figure 16.12b. That is a sudden process, not at all quasi-static. The *irreversible* process of Figure 16.12b cannot be represented on a pV diagram.

The critical question is: How slow must a process be to qualify as quasi-static? That turns out to be a difficult question to answer. This textbook will always assume that processes are quasi-static. It turns out to be a reasonable assumption for the types of examples and homework problems we will look at. Irreversible processes will be left to more advanced courses.

Constant-Volume Process

Many important gas processes take place in a container of constant, unchanging volume. A constant-volume process is called an **isochoric process,** where *iso* is a prefix meaning "constant" or "equal" while *choric* is from a Greek root meaning "volume." An isochoric process is one for which

$$V_f = V_i \tag{16.14}$$

For example, suppose that you have a gas in the closed, rigid container shown in Figure 16.13a. Warming the gas with a Bunsen burner will raise its pressure without changing its volume. This process is shown as the vertical line $1 \rightarrow 2$ on the pV diagram of Figure 16.13b. A constant-volume cooling, by placing the container on a block of ice, would lower the pressure and be represented as the vertical line from 2 to 1. **Any isochoric process appears on a pV diagram as a vertical line.**

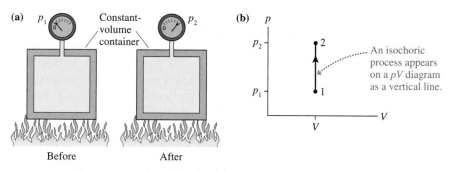

FIGURE 16.13 A constant-volume (isochoric) process.

EXAMPLE 16.6 A constant-volume gas thermometer

A constant-volume gas thermometer is placed in contact with a reference cell containing water at the triple point. After reaching equilibrium, the gas pressure is recorded as 55.78 kPa. The thermometer is then placed in contact with a sample of unknown temperature. After the thermometer reaches a new equilibrium, the gas pressure is 65.12 kPa. What is the temperature of this sample?

MODEL The thermometer's volume doesn't change, so this is an isochoric process.

SOLVE The temperature at the triple point of water is $T_1 = 0.01°C = 273.16$ K. The ideal-gas law for a closed system is $p_2 V_2/T_2 = p_1 V_1/T_1$. The volume doesn't change, so $V_2/V_1 = 1$. Thus

$$T_2 = T_1 \frac{V_2}{V_1} \frac{p_2}{p_1} = T_1 \frac{p_2}{p_1} = (273.16 \text{ K}) \frac{65.12 \text{ kPa}}{55.78 \text{ kPa}}$$

$$= 318.90 \text{ K} = 45.75°C$$

The temperature *must* be in kelvins to do this calculation, although it is common to convert the final answer to °C. The fact that the pressures were given to four significant figures justified using $T_K = T_C + 273.15$ rather than the usual $T_C + 273$.

ASSESS $T_2 > T_1$, which we expected from the increase in pressure.

Constant-Pressure Process

Other gas processes take place at a constant, unchanging pressure. A constant-pressure process is called an **isobaric process,** where *baric* is from the same root as "barometer" and means "pressure." An isobaric process is one for which

$$p_f = p_i \tag{16.15}$$

Figure 16.14a shows one method of changing the state of a gas while keeping the pressure constant. A cylinder of gas has a tight-fitting, massless piston that can slide up and down but seals the container so that no atoms enter or escape. The mass on top applies a constant downward force Mg on the piston. The air also presses down on the piston. In equilibrium, the upward force pA on the piston by the gas in the cylinder, where A is the piston's area, exactly balances the downward force Mg plus the air-pressure force $p_{atmos}A$. Consequently, the gas pressure inside the cylinder is

$$p = p_{atmos} + \frac{Mg}{A} \tag{16.16}$$

In other words, the gas pressure is determined by the requirement that the gas must support both the mass on top of the piston and the air pressing inward. This pressure is independent of the temperature of the gas or the height of the piston, so it stays constant as long as M is unchanged.

If the cylinder is warmed, the gas will expand and push the piston up. But the pressure, determined by mass M, will not change. This process is shown on the pV diagram of Figure 16.14b as the horizontal line $1 \rightarrow 2$. We call this an *isobaric expansion.* An *isobaric compression* occurs if the gas is cooled, lowering the piston. **Any isobaric process appears on a pV diagram as a horizontal line.**

(a) The mass maintains a constant pressure in the cylinder.

Before   After

(b) An isobaric process appears on a pV diagram as a horizontal line.

FIGURE 16.14 A constant-pressure (isobaric) process.

EXAMPLE 16.7 A constant-pressure compression
A gas occupying 50 cm³ at 50°C is cooled at constant pressure until the temperature is 10° C. What is its final volume?

MODEL The pressure of the gas doesn't change, so this is an isobaric process.

SOLVE By definition, $p_1/p_2 = 1$ for an isobaric process. Using the ideal-gas law for constant n, we have

$$V_2 = V_1 \frac{p_1}{p_2} \frac{T_2}{T_1} = V_1 \frac{T_2}{T_1}$$

Temperatures *must* be in kelvins to use the ideal-gas law. Thus

$$V_2 = (50 \text{ cm}^3) \frac{(10 + 273) \text{ K}}{(50 + 273) \text{ K}} = 43.8 \text{ cm}^3$$

ASSESS As long as we use *ratios,* we do not need to convert volumes or pressures to SI units. That is because the conversion is a multiplicative factor that cancels. But the conversion of temperature is an *additive* factor that does *not* cancel. That is why you must always convert temperatures to kelvins in ideal-gas calculations.

Constant-Temperature Process

The last process we wish to look at for now is one that takes place at a constant temperature. A constant-temperature process is called an **isothermal process.** An isothermal process is one for which $T_f = T_i$. Because $pV = nRT$, a constant-

Actjv Physics   8.4

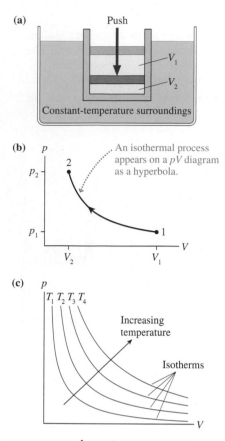

(a)

Push

V_1

V_2

Constant-temperature surroundings

(b)

An isothermal process appears on a pV diagram as a hyperbola.

FIGURE 16.15 A constant-temperature (isothermal) process.

temperature process in a closed system (constant n) is one for which the product pV doesn't change. Thus

$$p_f V_f = p_i V_i \qquad (16.17)$$

in an isothermal process.

One possible isothermal process is illustrated in Figure 16.15a. A piston is being pushed down to compress a gas, but the gas cylinder is floating in a large container of liquid that is held at a constant temperature. If the piston is pushed *slowly,* then heat energy transfer through the walls of the cylinder will keep the gas at the same temperature as the surrounding liquid. This would be an *isothermal compression.* The reverse process, with the piston slowly pulled out, would be an *isothermal expansion.*

Representing an isothermal process on the pV diagram is a little more complicated than the two previous processes because both p and V change. As long as T remains fixed, we have the relationship

$$p = \frac{nRT}{V} = \frac{\text{constant}}{V} \qquad (16.18)$$

The inverse relationship between p and V causes the graph of an isothermal process to be a *hyperbola.*

The process shown as $1 \rightarrow 2$ in Figure 16.15b represents the *isothermal compression* shown in Figure 16.15a. An *isothermal expansion* would move in the opposite direction along the hyperbola.

The location of the hyperbola depends on the value of T. A lower-temperature process is represented by a hyperbola closer to the origin than a higher-temperature process. Figure 16.15c shows four hyperbolas representing the temperatures T_1 to T_4 where $T_4 > T_3 > T_2 > T_1$. These are called **isotherms.** A gas undergoing an isothermal process will move along the isotherm of the appropriate temperature.

EXAMPLE 16.8 A constant-temperature compression

A gas cylinder with a tight-fitting, moveable piston contains 200 cm^3 of air at 1.0 atm. It floats on the surface of a swimming pool of 15°C water. The cylinder is then slowly pulled underwater to a depth of 3.0 m. What is the volume of gas at this depth?

MODEL The gas's temperature doesn't change, so this is an isothermal compression.

SOLVE At the surface, the pressure inside the cylinder must exactly equal the outside air pressure of 1.0 atm. If the pressures were not equal, a net force would push the piston in or pull it out until the pressures balanced and equilibrium was achieved. As the cylinder is pulled underwater, the increasing water pressure pushes the piston in and compresses the gas. Equilibrium at depth d requires that the gas pressure inside the cylinder equal

the water pressure $p_{water} = p_0 + \rho g d$, where $p_0 = 1.0$ atm is the pressure at the surface. As long as the cylinder moves slowly, the gas will stay at the same temperature as the surrounding water. The value of T is not important; all we need to know is that the compression is isothermal. In that case, because $T_2/T_1 = 1$,

$$V_2 = V_1 \frac{T_2}{T_1} \frac{p_1}{p_2} = V_1 \frac{p_1}{p_2} = V_1 \frac{p_0}{p_0 + \rho g d}$$

The initial pressure p_0 must be in SI units: $p_0 = 1.0$ atm $= 1.013 \times 10^5$ Pa. Then a straightforward computation gives $V_2 = 155 \text{ cm}^3$.

ASSESS V_2 is less than V_1. This is expected because the gas is being compressed.

EXAMPLE 16.9 A multistep process

A gas at 2.0 atm pressure and a temperature of 200°C is first expanded isothermally until its volume has doubled. It then undergoes an isobaric compression until it returns to its original volume. First show this process on a pV diagram. Then find the final temperature and pressure.

MODEL Many practical ideal-gas processes consists of several basic steps performed in series. In this case, the final state of the isothermal expansion is the initial state for an isobaric compression.

VISUALIZE Figure 16.16 shows the process. The gas starts in state 1 at pressure $p_1 = 2.0$ atm and volume V_1. As the gas

FIGURE 16.16 A pV diagram for the process of Example 16.9.

at constant pressure p_2 until its final volume V_3 equals its original volume V_1. State 3 is on an isotherm closer to the origin, so we expect to find $T_3 < T_1$.

SOLVE $T_2/T_1 = 1$ during the isothermal expansion and $V_2 = 2V_1$, so the pressure at point 2 is

$$p_2 = p_1 \frac{T_2}{T_1} \frac{V_1}{V_2} = p_1 \frac{V_1}{2V_1} = \frac{1}{2} p_1 = 1.0 \text{ atm}$$

$p_3/p_2 = 1$ during the isobaric compression and $V_3 = V_1 = \frac{1}{2} V_2$, so

$$T_3 = T_2 \frac{p_3}{p_2} \frac{V_3}{V_2} = T_2 \frac{\frac{1}{2} V_2}{V_2} = \frac{1}{2} T_2 = 236.5 \text{ K} = -36.5°\text{C}$$

where we converted T_2 to 473 K before doing calculations and then converted T_3 back to °C. The final state, with $T_3 = -36.5°\text{C}$ and $p_3 = 1.0$ atm, is one in which both the pressure and the absolute temperature are half their original values.

expands isothermally, it moves downward along an isotherm until it reaches volume $V_2 = 2V_1$. The pressure decreases during this process to a lower value, p_2. The gas is then compressed

---

STOP TO THINK 16.6 What is the ratio T_f/T_i for this process?

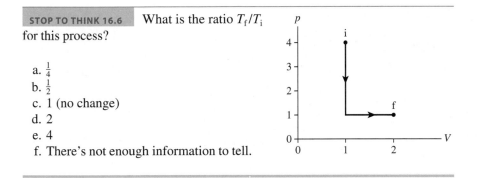

a. $\frac{1}{4}$
b. $\frac{1}{2}$
c. 1 (no change)
d. 2
e. 4
f. There's not enough information to tell.

SUMMARY

The goal of Chapter 16 has been to learn the characteristics of macroscopic systems.

GENERAL PRINCIPLES

Three Phases of Matter

Solid   Rigid, definite shape. Nearly incompressible.

Liquid   Molecules loosely held together by molecular bonds, but able to move around. Nearly incompressible.

Gas   Molecules move freely through space. Compressible.

The different phases exist for different conditions of temperature T and pressure p. The boundaries separating the regions of a phase diagram are lines of phase equilibrium. Any amounts of the two phases can coexist in equilibrium. The **triple point** is the one value of temperature and pressure at which all three phases can coexist in equilibrium.

IMPORTANT CONCEPTS

Ideal-Gas Model

- Atoms and molecules are small, hard spheres that travel freely through space except for occasional collisions with each other or the walls.

- The molecules have a distribution of speeds.

- The model is valid when the density is low and the temperature well above the condensation point.

Ideal-Gas Law

The **state variables** of an ideal gas are related by the ideal-gas law

$$pV = nRT \quad \text{or} \quad pV = Nk_BT$$

where $R = 8.31$ J/mol K is the universal gas constant and $k_B = 1.38 \times 10^{-23}$ J/K is Boltzmann's constant.

p, V, and T *must* be in SI units of Pa, m^3, and K. For a gas in a sealed container, with constant n:

$$\frac{p_2V_2}{T_2} = \frac{p_1V_1}{T_1}$$

Counting atoms and moles

A macroscopic sample of matter consists of N atoms (or molecules), each of mass m (the **atomic** or **molecular mass**):

$$N = \frac{M}{m}$$

Alternatively, we can state that the sample consists of n **moles**

$$n = \frac{N}{N_A} \quad \text{or} \quad \frac{M(\text{in grams})}{M_{mol}}$$

$N_A = 6.02 \times 10^{23}$ mol^{-1} is **Avogadro's number.**

The numerical value of the molar mass M_{mol}, in g/mol, equals the numerical value of the atomic or molecular mass m in u. The atomic or molecular mass m, in atomic mass units u, is well approximated by the **atomic mass number A.**

$$1 \text{ u} = 1.661 \times 10^{-27} \text{ kg}$$

The **number density** of the sample is $\dfrac{N}{V}$.

APPLICATIONS

Temperature scales

$$T_F = \frac{9}{5}T_C + 32° \qquad T_K = T_C + 273$$

The Kelvin temperature scale is based on:

- Absolute zero at $T_0 = 0$ K

- The triple point of water at $T_3 = 273.16$ K

Three basic gas processes

1. **Isochoric,** or constant volume
2. **Isobaric,** or constant pressure
3. **Isothermal,** or constant temperature

pV diagram

TERMS AND NOTATION

macroscopic system	atomic mass number, A	absolute temperature scale	universal gas constant, R
bulk properties	atomic mass	melting point	Boltzmann's constant, k_B
micro/macro connection	atomic mass unit, u	freezing point	STP
phase	molecular mass	phase equilibrium	ideal-gas process
phase change	mole, n	condensation point	pV diagram
solid	monatomic gas	boiling point	quasi-static process
crystal	diatomic gas	phase diagram	isochoric process
amorphous solid	Avogadro's number, N_A	sublimation	isobaric process
liquid	molar mass, M_{mol}	critical point	isothermal process
gas	temperature, T	triple point	isotherm
state variable	constant-volume gas	ideal gas	
thermal equilibrium	thermometer	histogram	
number density, N/V	absolute zero, T_0	ideal-gas law	

EXERCISES AND PROBLEMS

Exercises

Section 16.1 Solids, Liquids, and Gases

1. What volume of water has the same mass as 2.0 m³ of lead?
2. The nucleus of a uranium atom has a diameter of 1.5×10^{-14} m and a mass of 4.0×10^{-25} kg. What is the density of the nucleus?
3. What is the diameter of a copper sphere that has the same mass as a 10 cm × 10 cm × 10 cm cube of aluminum?
4. A hollow sphere, with an outer diameter of 10.0 cm and an inner diameter of 8.0 cm, has a mass of 690 g. Of what material is it made?

Section 16.2 Atoms and Moles

5. How many atoms are in a 2.0 cm × 2.0 cm × 2.0 cm cube of aluminum?
6. How many moles are in a 2.0 cm × 2.0 cm × 2.0 cm cube of copper?
7. What is the number density of (a) aluminum and (b) lead?
8. An element in its solid phase has mass density 1750 kg/m³ and number density 4.39×10^{28} atoms/m³. What is the element's atomic mass number?
9. 1.0 mol of gold is shaped into a cube. What is the length of each edge?
10. What volume of aluminum has the same number of atoms as 10 cm³ of mercury?

Section 16.3 Temperature

11. The lowest and highest natural temperatures ever recorded on earth are $-127°F$ in Antarctica and $136°F$ in Libya. What are these temperatures in °C and in K?
12. At what temperature does the numerical value in °F match the numerical value in °C?

13. A demented scientist creates a new temperature scale, the "Z scale." He decides to call the boiling point of nitrogen 0°Z and the melting point of steel 1000°Z.
    a. What is the boiling point of water on the Z scale?
    b. Convert 500°Z to the Celsius scale and to the Kelvin scale.
14. The sample in an experiment is initially at 10°C. If the sample's temperature is doubled, what is the new temperature in °C?

Section 16.4 Phase Changes

15. What is the temperature in °F and the pressure in Pa at the triple point of (a) water and (b) carbon dioxide?
16. a. Is there a highest temperature at which ice can exist? If so, what is it? If not, why not?
    b. Is there a lowest temperature at which water vapor can exist? If so, what is it? If not, why not?
17. An aquanaut lives in an underwater apartment 100 m beneath the surface of the ocean. Compare the freezing and boiling points of water in the aquanaut's apartment to their values at the surface. Are they higher, lower, or the same? Explain.
18. a. A sample of water vapor in an enclosed cylinder has an initial pressure of 500 Pa at an initial temperature of $-0.01°C$. A piston squeezes the sample smaller and smaller, without limit. Describe what happens to the water as the squeezing progresses.
    b. Repeat part a if the initial temperature is 0.03°C warmer.

Section 16.5 Ideal Gases

19. A rigid container holds 2.0 mol of gas at a pressure of 1.0 atm and a temperature of 30°C.
    a. What is the container's volume?
    b. What is the pressure if the temperature is raised to 130°C?

20. A cylinder contains 4.0 g of nitrogen gas. A piston compresses the gas to half its initial volume. Afterward,
    a. Has the mass density of the gas changed? If so, by what factor? If not, why not?
    b. Has the number of moles of gas changed? If so, by what factor? If not, why not?
21. 3.0 mol of gas at a temperature of $-120°C$ fills a 2.0 L container. What is the gas pressure?
22. A gas at $100°C$ fills volume V_0. If the pressure is held constant, what is the volume if (a) the Celsius temperature is doubled and (b) the Kelvin temperature is doubled.
23. A 15-cm-diameter compressed-air tank is 50 cm tall. The pressure at $20°C$ is 150 atm.
    a. How many moles of air are in the tank?
    b. What volume would this air occupy at STP?
24. A 20-cm-diameter cylinder that is 40 cm long contains 50 g of oxygen gas at $20°C$.
    a. How many moles of oxygen are in the cylinder?
    b. How many oxygen molecules are in the cylinder?
    c. What is the number density of the oxygen?
    d. What is the reading of a pressure gauge attached to the tank?
25. A 10-cm-diameter cylinder of helium gas is 30 cm long and at $20°C$. The pressure gauge reads 120 psi.
    a. How many helium atoms are in the cylinder?
    b. What is the mass of the helium?
    c. What is the helium number density?
    d. What is the helium mass density?
26. A 1.0 m $\times$ 1.0 m $\times$ 1.0 m cube of nitrogen gas is at $20°C$ and 1.0 atm. Estimate the number of molecules in the cube with a speed between 700 m/s and 1000 m/s.

Section 16.6 Ideal-Gas Processes

27. A rigid sphere has a valve that can be opened or closed. The sphere with the valve open is placed in boiling water in a room where the air pressure is 1.0 atm. After a long period of time has elapsed, the valve is closed. What will be the pressure inside the sphere if it is then placed in (a) a mixture of ice and water and (b) an insulated box filled with dry ice?
28. A gas with initial state variables p_1, V_1, and T_1 expands isothermally until $V_2 = 2V_1$. What are (a) T_2 and (b) p_2?
29. 0.10 mol of argon gas is admitted to an evacuated 50 cm^3 container at $20°C$. The gas then undergoes an isochoric heating to a temperature of $300°C$.
    a. What is the final pressure of the gas?
    b. Show the process on a pV diagram. Include a proper scale on both axes.
30. 0.10 mol of argon gas is admitted to an evacuated 50 cm^3 container at $20°C$. The gas then undergoes an isobaric heating to a temperature of $300°C$.
    a. What is the final volume of the gas?
    b. Show the process on a pV diagram. Include a proper scale on both axes.
31. 0.10 mol of argon gas is admitted to an evacuated 50 cm^3 container at $20°C$. The gas then undergoes an isothermal expansion to a volume of 200 cm^3.
    a. What is the final pressure of the gas?
    b. Show the process on a pV diagram. Include a proper scale on both axes.

32. 0.0040 mol of gas undergoes the process shown in Figure Ex16.32.
    a. What type of process is this?
    b. What are the initial and final temperatures?

FIGURE EX16.32

33. 0.0040 mol of gas undergoes the process shown in Figure Ex16.33.
    a. What type of process is this?
    b. What are the initial and final temperatures?
    c. What is the final volume V_2?

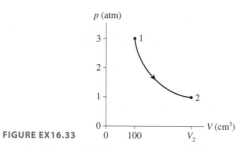

FIGURE EX16.33

34. A gas with an initial temperature of $900°C$ undergoes the process shown in Figure Ex16.34.
    a. What type of process is this?
    b. What is the final temperature?
    c. How many moles of gas are there?

FIGURE EX16.34

Problems

35. The atomic mass number of copper is $A = 64$. Assume that atoms in solid copper form a cubic crystal lattice. To envision this, imagine that you place atoms at the centers of tiny sugar cubes, then stack the little sugar cubes to form a big cube. If you dissolve the sugar, the atoms left behind are in a cubic crystal lattice. What is the smallest distance between two copper atoms?
36. An element in its solid phase forms a cubic crystal lattice (see Problem 35) with mass density 7950 kg/m^3. The smallest spacing between two adjacent atoms is 0.227 nm. What is the element's atomic mass number?

37. a. Use data from Figure 16.8 to estimate the average kinetic energy of a nitrogen molecule at room temperature.
    b. You will learn in Chapter 18 that the average molecular kinetic energy at temperature T is *independent* of the composition of the gas. Assuming this, estimate the average speed of a hydrogen molecule at room temperature.

38. Estimate the number density of gas molecules in the earth's atmosphere at sea level.

39. The solar corona is a very hot atmosphere surrounding the visible surface of the sun. X-ray emissions from the corona show that its temperature is about 2×10^6 K. The gas pressure in the corona is about 0.03 Pa. Estimate the number density of particles in the solar corona.

40. Current vacuum technology can achieve a pressure of 1.0×10^{-10} mm of Hg. At this pressure, and at a temperature of 20°C, how many molecules are in 1 cm³?

41. The semiconductor industry manufactures integrated circuits in large vacuum chambers where the pressure is 1.0×10^{-10} mm of Hg.
    a. What fraction is this of atmospheric pressure?
    b. At $T = 20$°C, how many molecules are in a cylindrical chamber 40 cm in diameter and 30 cm tall?

42. On average, each person in the industrialized world is responsible for the emission of 10,000 kg of carbon dioxide (CO_2) every year. This includes CO_2 that you generate directly, by burning fossil fuels to operate your car or your furnace, as well as CO_2 generated on your behalf by electric generating stations and manufacturing plants. CO_2 is a greenhouse gas that contributes to global warming. If you were to store your yearly CO_2 emissions in a cube at STP, how long would each edge of the cube be?

43. A gas at 25°C and atmospheric pressure fills a cylinder. The gas is transferred to a new cylinder with three times the volume, after which the pressure is half the original pressure. What is the new temperature of the gas?

44. What is the ratio of the volume occupied by 300°C steam at 1.0 atm pressure to the volume occupied by an equal mass of liquid water?

45. 10,000 cm³ of 200°C steam at a pressure of 20 atm is cooled until it condenses. What is the volume of the liquid water? Give your answer in cm³.

46. An electric generating plant boils water to produce high-pressure steam. The steam spins a turbine that is connected to the generator.
    a. How many liters of water must be boiled to fill a 5.0 m³ boiler with 50 atm of steam at 400°C?
    b. The steam has dropped to 2.0 atm pressure at 150°C as it exits the turbine. How much volume does it now occupy?

47. On a cool morning, when the temperature is 15°C, you measure the pressure in your car tires to be 30 psi. After driving 20 mi on the freeway, the temperature of your tires is 45°C. What pressure will your tire gauge now show?

48. The air temperature and pressure in a laboratory are 20°C and 1.0 atm. A 1.0 L container is open to the air. The container is then sealed and placed in a bath of boiling water. After reaching thermal equilibrium, the container is opened. How many moles of air escape?

49. The volume in a constant-*pressure* gas thermometer is directly proportional to the absolute temperature. A constant-pressure thermometer is calibrated by adjusting its volume to 1000 mL while it is in contact with a reference cell at the triple point of water. The volume increases to 1638 mL when the thermometer is placed in contact with a sample. What is the sample's temperature?

50. The mercury manometer shown in Figure P16.50 is attached to a gas cell. The mercury height h is 120 mm when the cell is placed in an ice-water mixture. The mercury height drops to 30 mm when the device is carried into an industrial freezer. What is the freezer temperature?
    Hint: The right tube of the manometer is much narrower than the left tube. What reasonable assumption can you make about the gas volume?

FIGURE P16.50

51. A 200 cm³ gas cell containing helium at 50°C is connected to a mercury manometer. The mercury in the outside arm of the manometer is 25.0 cm below the mercury in the arm attached to the gas cell.
    a. How many helium atoms are in the gas cell?
    b. What is the mass of the helium?

52. A diver 50 m deep in 10°C fresh water exhales a 1.0-cm-diameter bubble. What is the bubble's diameter just as it reaches the surface of the lake, where the water temperature is 20°C?
    Hint: Assume that the air bubble is always in thermal equilibrium with the surrounding water.

53. A compressed-air cylinder is known to fail if the pressure exceeds 110 atm. A cylinder that was filled to 25 atm at 20°C is stored in a warehouse. Unfortunately, the warehouse catches fire and the temperature reaches 950°C. Does the cylinder blow?

54. Reproduce Figure P16.54 on a piece of paper. A gas starts with pressure p_1 and volume V_1. Show on the figure the process in which the gas undergoes an isochoric process that doubles the pressure, then an isobaric process that doubles the volume, followed by an isothermal process that doubles the volume again. Label each of the three processes.

FIGURE P16.54   **FIGURE P16.55**

55. Reproduce Figure P16.55 on a piece of paper. A gas starts with pressure p_1 and volume V_1. Show on the figure the process in which the gas undergoes an isothermal process during which the volume is halved, then an isochoric process during which the pressure is halved, followed by an isobaric process during which the volume is doubled. Label each of the three processes.

56. 8.0 g of helium gas follows the process $1 \rightarrow 2 \rightarrow 3$ shown in Figure P16.56. Find the values of V_1, V_3, p_2, and T_3.

FIGURE P16.56   **FIGURE P16.57**

57. Figure P16.57 shows two different processes by which 1.0 g of nitrogen gas moves from state 1 to state 2. The temperature of state 1 is 25°C. What are (a) pressure p_1 and (b) temperatures (in °C) T_2, T_3, and T_4?

58. Figure P16.58 shows two different processes by which 80 mol of gas move from state 1 to state 2. The dashed line is an isotherm.
    a. What is the temperature of the isothermal process?
    b. What maximum temperature is reached along the straight-line process?

FIGURE P16.58   **FIGURE P16.59**

59. 0.10 mol of gas undergoes the process $1 \rightarrow 2$ shown in Figure P16.59.
    a. What are temperatures T_1 and T_2?
    b. What type of process is this?
    c. The gas undergoes an isochoric heating from point 2 until the pressure is restored to the value it had at point 1. What is the final temperature of the gas?

60. 0.0050 mol of gas undergoes the process $1 \rightarrow 2 \rightarrow 3$ shown in Figure P16.60. What are (a) temperature T_1, (b) pressure p_2, and (c) volume V_3?

FIGURE P16.60

61. 10 g of dry ice (solid CO_2) is placed in a 10,000 cm³ container, then all the air is quickly pumped out and the container sealed. The container is warmed to 0°C, a temperature at which CO_2 is a gas.
    a. What is the gas pressure? Give your answer in atm.
       The gas then undergoes an isothermal compression until the pressure is 3.0 atm, immediately followed by an isobaric compression until the volume is 1000 cm³.

    b. What is the final temperature of the gas?
    c. Show the process on a pV diagram.

62. A container with 100 cm³ of gas is pressurized at constant volume until the gas pressure triples. It is then expanded at constant temperature until the pressure is one-half of the pressure the gas had before the experiment started.
    a. What is the final volume of the gas?
    b. Show this process on a pV diagram.

63. A container of gas at 2.0 atm pressure and 127°C is compressed at constant temperature until the volume is halved. It is then further compressed at constant pressure until the volume is halved again.
    a. What are the final pressure and temperature of the gas?
    b. Show this process on a pV diagram.

64. Five grams of nitrogen gas at an initial pressure of 3.0 atm and at 20°C undergo an isobaric expansion until the volume has tripled.
    a. What is the gas volume after the expansion?
    b. What is the gas temperature after the expansion?
       The gas pressure is then decreased at constant volume until the original temperature is reached.
    c. What is the gas pressure after the decrease?
       Finally, the gas is isothermally compressed until it returns to its initial volume.
    d. What is the final gas pressure?
    e. Show the full three-step process on a pV diagram. Use appropriate scales on both axes.

In Problems 65 through 68 you are given the equation(s) used to solve a problem. For each of these, you are to
   a. Write a realistic problem for which this is the correct equation(s).
   b. Draw a pV diagram.
   c. Finish the solution of the problem.

65. $p_2 = \dfrac{300 \text{ cm}^3}{100 \text{ cm}^3} \times 1 \times 2 \text{ atm}$

66. $(T_2 + 273) \text{ K} = \dfrac{200 \text{ kPa}}{500 \text{ kPa}} \times 1 \times (400 + 273) \text{ K}$

67. $V_2 = \dfrac{(400 + 273) \text{ K}}{(50 + 273) \text{ K}} \times 1 \times 200 \text{ cm}^3$

68. $(2.0 \times 101{,}300 \text{ Pa})(100 \times 10^{-6} \text{ m}^3) = n(8.31 \text{ J/mol K})T_1$

$n = \dfrac{0.12 \text{ g}}{20 \text{ g/mol}}$

$T_2 = \dfrac{200 \text{ cm}^3}{100 \text{ cm}^3} \times 1 \times T_1$

Challenge Problems

69. The 50 kg lead piston shown in Figure CP16.69 floats on 0.12 mol of compressed air.
    a. What is the piston height h if the temperature is 30°C?
    b. How far does the piston move if the temperature is increased by 100°C?

FIGURE CP16.69

70. The U-shaped tube in Figure CP16.70 has a total length of 1.0 m. It is open at one end, closed at the other, and is initially filled with air at 20°C and 1.0 atm pressure. Mercury is poured slowly into the open end without letting any air escape, thus compressing the air. This is continued until the open side of the tube is completely filled with mercury. What is the length L of the column of mercury?

FIGURE CP16.70

71. A diving bell is a 3.0-m-tall cylinder closed at the upper end but open at the lower end. The temperature of the air in the bell is 20°C. The bell is lowered into the ocean until its lower end is 100 m deep. The temperature at that depth is 10°C.
    a. How high does the water rise in the bell after enough time has passed for the air to reach thermal equilibrium?
    b. A compressed-air hose from the surface is used to expel all the water from the bell. What minimum air pressure is needed to do this?

72. The 3.0-m-long pipe in Figure CP16.72 is closed at the top end. It is slowly pushed straight down into the water until the top end of the pipe is level with the water's surface. What is the length L of the trapped volume of air?

FIGURE CP16.72

73. The cylinder in Figure CP16.73 has a moveable piston attached to a spring. The cylinder's cross-section area is 10 cm², it contains 0.0040 mol of gas, and the spring constant is 1500 N/m. At 20°C the spring is neither compressed nor stretched. How far is the spring compressed if the gas temperature is raised to 100°C?

FIGURE CP16.73

74. Containers A and B in Figure CP16.74 hold the same gas. The volume of B is four times the volume of A. The two containers are connected by a thin tube (negligible volume) and a valve that is closed. The gas in A is at 300 K and pressure of 1.0×10^5 Pa. The gas in B is at 400 K and pressure of 5.0×10^5 Pa. Heaters will maintain the temperatures of A and B even after the valve is opened.
    a. After the valve is opened, gas will flow one way or the other until A and B have equal pressure. What is this final pressure?
    b. Is this a reversible or an irreversible process? Explain.

FIGURE CP16.74

STOP TO THINK ANSWERS

Stop to Think 16.1: d. The pressure *decreases* by 20 kPa.

Stop to Think 16.2: a. The number of atoms depends only on the number of moles, not the substance.

Stop to Think 16.3: a. The step size on the Kelvin scale is the same as the step size on the Celsius scale. A *change* of 10°C is a *change* of 10 K.

Stop to Think 16.4: a. On the water phase diagram, you can see that for a pressure just slightly below the triple-point pressure that the solid/gas transition occurs at a higher temperature than does the solid/liquid transition at high pressures. This is not true for carbon dioxide.

Stop to Think 16.5: c. $T = pV/nR$. Pressure and volume are the same, but n differs. The number of moles in mass M is $n = M/M_{mol}$. Helium, with the smaller molar mass, has a larger number of moles and thus a lower temperature.

Stop to Think 16.6: b. The temperature decreases by a factor of 4 during the isochoric process, where $p_f/p_i = \frac{1}{4}$. The temperature then increases by a factor of 2 during the isobaric expansion, where $V_f/V_i = 2$.

17 Work, Heat, and the First Law of Thermodynamics

The modern technological era was ushered in by a new understanding of the connection between work and heat.

▶ **Looking Ahead**

The goal of Chapter 17 is to expand our understanding of energy and to develop the first law of thermodynamics as a general statement of energy conservation. In this chapter you will learn to:

- Understand the energy transfers known as *work* and *heat*.
- Use the first law of thermodynamics.
- Calculate work and heat for ideal-gas processes.
- Use specific heats and heats of transformation in the practical application of calorimetry.
- Understand adiabatic processes.

◀ **Looking Back**

The material in this chapter continues the development of energy ideas from Chapter 11. Many of the examples depend on the properties of ideal gases. Please review:

- Section 11.4 Work.
- Sections 11.7–11.8 Conservation of energy.
- Sections 16.4–16.6 Phase changes and ideal gases.

The industrial revolution was powered by the steam engine. Heat from a wood or coal fire was used to boil water and produce high-pressure steam. The expanding steam pushed a piston that, through a series of gears and levers, turned paddle wheels, ran machinery, or even powered massive locomotives. Humans had used heat for thousands of years for activities ranging from cooking to metallurgy, but the steam engine marked the first time in human history that heat was used to do work.

Our goal in this chapter is to investigate the connection between work and heat in macroscopic systems. Work and heat are *energy transfers* between the system and its environment, so we will be continuing the development of energy concepts that we began in Chapters 10 and 11. In addition, we will want to understand how the state of a system *changes* in response to work and heat. These two ideas, the transfer of energy and the change in the system, are related to each other through the *first law of thermodynamics,* a powerful statement about energy conservation.

We have two primary tasks: first, to complete our ideas about energy and its transfer; second, to investigate the consequences of energy transfer to and from macroscopic systems. Many practical devices, ranging from the internal combustion engine in your car to nuclear reactors, depend upon these consequences.

17.1 It's All About Energy

A key idea of Chapter 11 was the work-kinetic energy theorem in the form

$$\Delta K = W_c + W_{diss} + W_{ext} \tag{17.1}$$

Equation 17.1 tells us that the kinetic energy of a particle or a system of particles is changed when forces do work on the particles by pushing or pulling them through a distance. Here

1. W_c is the work done by conservative forces. This work can be represented as a change in the system's potential energy: $\Delta U = -W_c$.
2. W_{diss} is the work done by friction-like dissipative forces within the system. This work increases the system's thermal energy: $\Delta E_{th} = -W_{diss}$.
3. W_{ext} is the work done by external forces that originate in the environment. The tension in a rope would be an external force, as would the push of a piston rod.

With these definitions, Equation 17.1 becomes

$$\Delta K + \Delta U + \Delta E_{th} = W_{ext} \tag{17.2}$$

The system's *mechanical energy* was defined as $E_{mech} = K + U$. Figure 17.1 reminds you that the mechanical energy is associated with the motion of the system as a whole while E_{th} is associated with the motion of the atoms and molecules within the system. In the language of Part IV, E_{mech} is the *macroscopic* energy of the system as a whole while E_{th} is the *microscopic* energy of the particle-like atoms and spring-like molecular bonds. This led to our final energy statement of Chapter 11:

$$\Delta E_{sys} = \Delta E_{mech} + \Delta E_{th} = W_{ext} \tag{17.3}$$

where $E_{sys} = E_{mech} + E_{th}$ is the total energy of the system. We interpreted Equation 17.3 by saying that the total energy of an *isolated system,* one for which $W_{ext} = 0$, is constant. This was the essence of the law of conservation of energy as stated in Chapter 11.

The emphasis in Chapters 10 and 11 was on isolated systems, those with $W_{ext} = 0$. We were interested in learning how kinetic and potential energy were *transformed* into each other and, where there is friction, into thermal energy. Now we want to focus on how energy is *transferred* between the system and its environment. That is, on systems where W_{ext} is *not* zero. This section and the next will look more closely at thermal energy and work. Section 17.3 will then introduce *heat* as another form of energy transfer.

Thermal Energy

Thermal energy exists at the microscopic level of atoms and molecules. It is the sum of K_{micro}, the kinetic energy of all the moving atoms and molecules, and U_{micro}, the potential energy stored in the spring-like molecular bonds. That is,

$$E_{th} = K_{micro} + U_{micro} \tag{17.4}$$

The thermal energy is hidden from our macroscopic view but nonetheless is quite real.

The macroscopic energy of the system as a whole is its mechanical energy E_{mech}.

$\vec{v}$

$E_{sys} = E_{mech} + E_{th}$

The microscopic motion of the atoms and molecules is kinetic energy K_{micro}. The stretched and compressed molecular bonds have potential energy U_{micro}. Together, these are the system's thermal energy E_{th}.

FIGURE 17.1 The total energy of a system consists of the macroscopic mechanical energy of the system as a whole plus the microscopic thermal energy of the atoms.

Strictly speaking, the thermal energy due to molecular motion is only one form of energy that can be stored within a system at the microscopic level. For example, a system might have *chemical energy* that can be released via chemical reactions between molecules in the system. Chemical energy is quite important in engineering thermodynamics, where it is needed to characterize combustion processes. *Nuclear energy* is stored in the atomic nuclei and can be released during radioactive decay. All the sources of microscopic energy taken together are called the system's **internal energy:**

$$E_{int} = E_{th} + E_{chem} + E_{nuc} + \cdots \qquad (17.5)$$

The total energy of the system is then $E_{sys} = E_{mech} + E_{int}$. This textbook will concentrate on simple thermodynamic systems in which the internal energy is entirely thermal: $E_{int} = E_{th}$. We'll leave other forms of internal energy to more advanced courses.

Chapter 16 noted that E_{th} is associated with the system's temperature. We now need to be a bit more specific.

- During a phase change, from solid to liquid or from liquid to gas, a system's thermal energy increases but its temperature does not. We'll look at phase changes later in this chapter.
- If there are no phase changes, increasing the system's temperature increases its thermal energy. **Conversely, a system's thermal energy does not change ($\Delta E_{th} = 0$) during an isothermal process ($\Delta T = 0$).** This simple idea will have major consequences.

EXAMPLE 17.1 **The thermal energy of a gas**
Estimate the thermal energy of 1 mol of nitrogen gas at 20°C.

MODEL For a gas, the thermal energy is the total kinetic energy of the moving molecules. That is, $E_{th} = K_{micro}$.

SOLVE Chapter 16 looked at the distribution of speeds of N_2 molecules at 20°C (Figure 16.8). While there was a range of speeds, the average speed was $v_{avg} \approx 550$ m/s. Consequently, an average molecule in the gas has kinetic energy $K_{avg} = \frac{1}{2}mv_{avg}^2$. Nitrogen atoms have atomic mass number $A = 14$, so the mass of each molecule is

$$m = 28\,u = 4.65 \times 10^{-26}\,kg$$

Thus

$$K_{avg} = \frac{1}{2}mv_{avg}^2 \approx \frac{1}{2}(4.65 \times 10^{-26}\,kg)(550\,m/s)^2$$
$$\approx 7.0 \times 10^{-21}\,J$$

There are N_A molecules in 1 mol of gas, so the thermal energy is

$$E_{th} = K_{micro} = N_A K_{avg} \approx 4200\,J$$

ASSESS This is a substantial amount of energy, far more than most people would suspect.

The pistons in a car engine do work on the air-fuel mixture by compressing it.

Work

We introduced the idea of *work* in Chapter 11. Now, in preparation for the new concept of *heat,* we need to make our ideas about work more precise. **Work** is the energy transferred between a system and the environment when a net force acts on the system over a distance. The process itself is a **mechanical interaction,** meaning that the system and the environment interact via macroscopic pushes and pulls. Loosely speaking, we say that the environment (or a particular force from the environment) "does work" on the system. A system is in **mechanical equilibrium** with the environment or with another system if there is no net force on the system.

Figure 17.2 reminds you that work can be either positive or negative. **The sign of the work is *not* just an arbitrary convention, nor does it have anything to do with the choice of coordinate system.** The sign of the work tells us which way energy is being transferred.

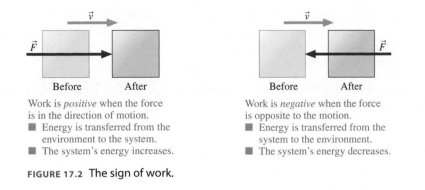

Work is *positive* when the force is in the direction of motion.
- Energy is transferred from the environment to the system.
- The system's energy increases.

Work is *negative* when the force is opposite to the motion.
- Energy is transferred from the system to the environment.
- The system's energy decreases.

FIGURE 17.2 The sign of work.

In contrast to the mechanical energy or the thermal energy, **work is not a state variable.** That is, work is not a number characterizing the system. Instead, work is the amount of energy that moves between the system and the environment during a mechanical interaction. We can measure the *change* in a state variable, such as a temperature change $\Delta T = T_f - T_i$, but it would make no sense to talk about a "change of work." Consequently, work appears in the energy equation simply as W_{ext}, never as ΔW_{ext}.

Energy Transfer

Doing work on a system can have very different consequences. Figure 17.3a shows an object being lifted at steady speed by a rope. The rope's tension is an external force doing work W_{ext} on the system. In this case, the energy transferred into the system goes entirely to increasing the system's macroscopic potential energy U_{grav}, part of the mechanical energy. The energy transfer process $W_{ext} \rightarrow E_{mech}$ is shown graphically in the energy bar chart of Figure 17.3a.

Contrast this with Figure 17.3b, where the same rope with the same tension now drags the object at steady speed across a rough surface. The tension does the same amount of work, but the mechanical energy does not change. Instead, friction increases the thermal energy of the object + surface system. The energy transfer process $W_{ext} \rightarrow E_{th}$ is shown in the energy bar chart of Figure 17.3b.

The point of this example is that the energy transferred to a system can go entirely to the system's mechanical energy, entirely to its thermal energy, or (imagine dragging the object up an incline) some combination of the two. The energy isn't lost, but where it ends up depends on the circumstances.

That Can't Be All

You can transfer energy into a system by the mechanical process of doing work on the system. But that can't be all there is to energy transfer. What happens when you place a pan of water on the stove and light the burner? The water temperature increases, so $\Delta E_{th} > 0$. But no work is done ($W_{ext} = 0$) and there is no change in the water's mechanical energy ($\Delta E_{mech} = 0$). This process clearly violates the energy equation $\Delta E_{mech} + \Delta E_{th} = W_{ext}$. What's wrong?

Nothing is wrong. The energy equation is correct as far as it goes, but it is incomplete. Work is energy transferred in a mechanical interaction, but that is not the only way a system can interact with its environment. Energy can also be transferred between the system and the environment if they have a *thermal interaction*. The energy transferred in a thermal interaction is called *heat*.

The symbol for heat is Q. When heat is included, the energy equation becomes

$$\Delta E_{sys} = \Delta E_{mech} + \Delta E_{th} = W + Q \tag{17.6}$$

Heat and work, now on an equal footing, are both energy transferred between the system and the environment.

(a)

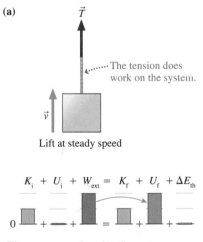

The tension does work on the system.

Lift at steady speed

$$K_i + U_i + W_{ext} = K_f + U_f + \Delta E_{th}$$

The energy transferred to the system goes entirely to the system's mechanical energy.

(b)

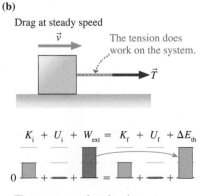

Drag at steady speed

The tension does work on the system.

$$K_i + U_i + W_{ext} = K_f + U_f + \Delta E_{th}$$

The energy transferred to the system goes entirely to the system's thermal energy.

FIGURE 17.3 The work done by the rope's tension can have very different consequences.

NOTE ▶ We've dropped the subscript "ext" from W. The work that we consider in thermodynamics is *always* the work done by the environment on the system. We won't need to distinguish this work from W_c or W_{diss}, so the subscript is superfluous. ◀

The next section looks at how work is calculated for ideal-gas processes. Heat will then be explored in Section 17.3.

STOP TO THINK 17.1 A gas cylinder and piston are covered with heavy insulation. The piston is pushed into the cylinder, compressing the gas. In this process, the gas temperature

a. increases.
b. decreases.
c. doesn't change.
d. there's not sufficient information to tell.

17.2 Work in Ideal-Gas Processes

You learned in Chapter 11 how to calculate work. The small amount of work dW done by force $\vec{F}$ as a system moves through the small displacement $d\vec{s}$ is $dW = \vec{F} \cdot d\vec{s}$. If we restrict ourselves to situations where $\vec{F}$ is either parallel or opposite to $d\vec{s}$, then the total work done on the system as it moves from s_i to s_f is

$$W = \int_{s_i}^{s_f} F_s \, ds \tag{17.7}$$

Work is positive if the displacement is in the same direction as the force, negative if the force and the displacement are in opposite directions.

Let's apply these definitions to a gas as it expands or is compressed. Figure 17.4a shows a gas cylinder that is sealed at one end by a movable piston. Force $\vec{F}_{ext}$, perhaps a force supplied by a piston rod, is equal in magnitude and opposite in direction to $\vec{F}_{gas}$. The gas pressure would blow the piston out of the cylinder if the external force weren't there! Using the coordinate system of Figure 17.4a,

$$F_{ext} = -F_{gas} = -pA \tag{17.8}$$

Suppose the piston moves the small distance dx shown in Figure 17.4b. As it does so, the external force (i.e., the environment) does work

$$dW = F_{ext} \, dx = -pA \, dx \tag{17.9}$$

If dx is positive (the gas expands), then dW is negative. This is because the external force is opposite the displacement. dW is positive if the gas is slightly compressed (negative dx) because the force and the displacement are in the same direction. This is an important idea.

As the piston moves dx, the volume of the gas changes by $dV = A \, dx$. Consequently, Equation 17.9 can be written in terms of the cylinder's volume as

$$dW = -p \, dV \tag{17.10}$$

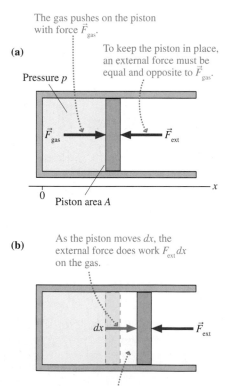

(a)

The gas pushes on the piston with force $\vec{F}_{gas}$.

Pressure p

To keep the piston in place, an external force must be equal and opposite to $\vec{F}_{gas}$.

$\vec{F}_{gas}$ $\vec{F}_{ext}$

0   Piston area A

(b)

As the piston moves dx, the external force does work $F_{ext} \, dx$ on the gas.

dx $\vec{F}_{ext}$

The volume changes by $dV = A \, dx$ as the piston moves dx.

FIGURE 17.4 The external force does work on the gas as the piston moves.

If we let the piston move in a slow quasi-static process from initial volume V_i to final volume V_f, the total work done by the environment on the gas is found by integrating Equation 17.10:

$$W = -\int_{V_i}^{V_f} p\,dV \qquad \text{(work done on a gas)} \qquad (17.11)$$

Equation 17.11 is a key result of thermodynamics. Although we used a cylinder to derive Equation 17.11, it turns out to be true for a container of any shape.

NOTE ▶ The pressure of a gas usually changes as the gas expands or contracts. Consequently, p is *not* a constant that can be brought outside the integral. You need to know how the pressure changes with volume before you can carry out the integration. ◀

We can give the work done on a gas a nice geometric interpretation. You learned in Chapter 16 how to represent an ideal-gas process as a curve in the pV diagram. Figure 17.5 shows that the work done on a gas is the negative of the area under the pV curve as the volume changes from V_i to V_f. That is

> W = the negative of the area under the pV curve between V_i and V_f

Figure 17.5a shows a process in which a gas *expands* from V_i to a larger volume V_f. The area under the curve is positive, so the environment does a negative amount of work on an expanding gas. Figure 17.5b shows a process in which a gas is compressed to a smaller volume. This one is a little trickier because we have to integrate "backward" along the V-axis. You learned in calculus that integrating from a larger limit to a smaller limit gives a negative result, so the area in Figure 17.5b is a negative area. Consequently, as the minus sign in Equation 17.11 indicates, the environment does a positive amount of work on a gas to compress it.

(a) For an *expanding* gas ($V_f > V_i$), the area under the pV curve is positive (integration direction is to the right). Thus the environment does *negative* work on the gas.

$W = -\text{area}$

Integration direction

(b) For a *compressed* gas ($V_f < V_i$), the area is negative because the integration direction is to the left. Thus the environment does *positive* work on the gas.

$W = -\text{area}$

Integration direction

FIGURE 17.5 The work done on a gas is the negative of the area under the curve.

EXAMPLE 17.2 The work done on an expanding gas

How much work is done on the gas in the ideal-gas process of Figure 17.6?

FIGURE 17.6 The ideal-gas process of Example 17.2.

MODEL The work done on a gas is the negative of the area under the pV curve. The gas is *expanding,* so we expect the work to be negative.

SOLVE The work W is the negative of the area under the curve from $V_i = 500$ cm³ to $V_f = 1500$ cm³. Volumes *must* be converted to SI units of m³. The area from 500 cm³ to 1000 cm³ can

be divided into a rectangle (between 0 kPa to 100 kPa) and a triangle (between 100 and 300 kPa). This area is

Area(500 → 1000 cm³)

$= ((1000 - 500) \times 10^{-6}\text{m}^3)(100{,}000\text{ Pa} - 0\text{ Pa})$

$+ \frac{1}{2}((1000 - 500) \times 10^{-6}\text{ m}^3)$

$\times (300{,}000\text{ Pa} - 100{,}000\text{ Pa})$

$= 100\text{ J}$

The area from 1000 cm³ to 1500 cm³ is a rectangle:

Area(1000 → 1500 cm³)

$= ((1500 - 1000) \times 10^{-6}\text{ m}^3)(300{,}000\text{ Pa} - 0\text{ Pa})$

$= 150\text{ J}$

The total area under the curve is 250 J, so the work done on the gas as it expands is

$$W = -(\text{area under the }pV\text{ curve}) = -250\text{ J}$$

ASSESS We noted previously that the product Pa m³ is equivalent to joules. The work is negative, as expected, because the external force pushing on the piston is opposite the direction of the piston's displacement.

Equation 17.11 is the basis for a problem-solving strategy.

PROBLEM-SOLVING STRATEGY 17.1 **Work in ideal-gas processes**

MODEL Assume the gas is ideal and the process is quasi-static.

VISUALIZE Show the process on a pV diagram. Note whether it happens to be one of the basic gas processes: isochoric, isobaric, or isothermal.

SOLVE Calculate the work as the area under the pV curve either geometrically or by carrying out the integration.

$$\text{Work done on the gas } W = -(\text{area under } pV \text{ curve}) = -\int p \, dV$$

ASSESS Check your signs.

■ $W > 0$ when the gas is compressed. Energy is transferred from the environment to the gas.
■ $W < 0$ when the gas expands. Energy is transferred from the gas to the environment.
■ No work is done if the volume doesn't change. $W = 0$.

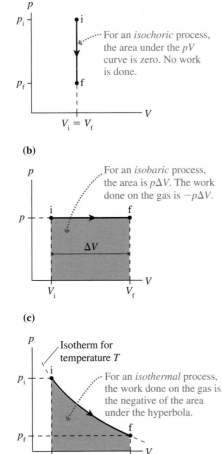

(a)

For an *isochoric* process, the area under the pV curve is zero. No work is done.

(b)

For an *isobaric* process, the area is $p\Delta V$. The work done on the gas is $-p\Delta V$.

(c)

Isotherm for temperature T

For an *isothermal* process, the work done on the gas is the negative of the area under the hyperbola.

FIGURE 17.7 Calculating the work done during ideal-gas processes.

Isochoric Process

The isochoric process in Figure 17.7a is one in which the volume does not change. Consequently,

$$W = 0 \qquad (\text{isochoric process}) \qquad (17.12)$$

An isochoric process is the *only* ideal-gas process in which no work is done.

Isobaric Process

Figure 17.7b shows an isobaric process in which the volume changes from V_i to V_f. The rectangular area under the curve is $p\Delta V$, so the work done during this process is

$$W = -p\Delta V \qquad (\text{isobaric process}) \qquad (17.13)$$

where $\Delta V = V_f - V_i$. ΔV is positive if the gas expands ($V_f > V_i$), so W is negative. ΔV is *negative* if the gas is compressed ($V_f < V_i$), making W positive.

Isothermal Process

Figure 17.7c shows an isothermal process. Here we need to know the pressure as a function of volume before we can integrate Equation 17.11. From the ideal-gas law, $p = nRT/V$. Thus the work on the gas as the volume changes from V_i to V_f is

$$W = -\int_{V_i}^{V_f} p \, dV = -\int_{V_i}^{V_f} \frac{nRT}{V} dV = -nRT \int_{V_i}^{V_f} \frac{dV}{V} \qquad (17.14)$$

where we could take the T outside the integral because temperature is constant during an isothermal process. This is a straightforward integration, giving

$$W = -nRT \int_{V_i}^{V_f} \frac{dV}{V} = -nRT \ln V \Big|_{V_i}^{V_f}$$

Activ
Physics   8.5
ONLINE

(17.15)

$$= -nRT(\ln V_f - \ln V_i) = -nRT \ln\left(\frac{V_f}{V_i}\right)$$

Because $nRT = p_i V_i = p_f V_f$ during an isothermal process, we can write the work three ways:

$$W = -nRT \ln\left(\frac{V_f}{V_i}\right) = -p_i V_i \ln\left(\frac{V_f}{V_i}\right) = -p_f V_f \ln\left(\frac{V_f}{V_i}\right)$$

(17.16)

$$\text{(isothermal process)}$$

Which version of Equation 17.16 is easiest to use will depend on the information you're given. The pressure, volume, and temperature *must* be in SI units.

EXAMPLE 17.3 The work of an isothermal compression
A cylinder contains 7.0 g of nitrogen gas. How much work must be done to compress the gas at a constant temperature of 80°C until the volume is halved?

MODEL This is an isothermal ideal-gas process.

SOLVE Nitrogen gas is N_2, with molar mass $M_{mol} = 28$ g/mol, so 7.0 g is 0.25 mol of gas. The temperature is $T = 353$ K. Although we don't know the actual volume, we do know that $V_f = \frac{1}{2}V_i$. The volume ratio is all we need to calculate the work:

$$W = -nRT \ln\left(\frac{V_f}{V_i}\right)$$
$$= -(0.25 \text{ mol})(8.31 \text{ J/mol K})(353 \text{ K})\ln\left(\frac{1}{2}\right) = 508 \text{ J}$$

ASSESS The work is positive because a force from the environment pushes the piston inward to compress the gas.

Work Depends on the Path

Figure 17.8 shows two different processes that take a gas from an initial state i to a final state f. Although the initial and final states are the same, the work done during these two processes is *not* the same. The area under the curve of process A is larger than the area under the curve of process B, so $|W_A| > |W_B|$. (The work is negative, so W_A is more negative than W_B.) As this example shows, **the work done during an ideal-gas process depends on the path followed through the pV diagram.**

FIGURE 17.8 The work done during these two ideal-gas processes is not the same.

You may be thinking that work is supposed to be independent of the path, but there are two big reasons why that is not the case here. First, we found in Chapter 11 that work is independent of the path only for a *conservative* force.

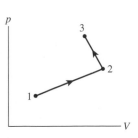

FIGURE 17.9 The work done during the process $1 \rightarrow 2 \rightarrow 3$ must be calculated in two steps.

In general, the forces that push and pull on pistons are not conservative forces. Second, the path that we considered in Chapter 11 was the trajectory of a particle from one point to another through space. For an ideal-gas process, the "path" is a sequence of thermodynamic states as they are represented on a pV diagram. It is a figurative path, because we can draw a picture of it on a pV diagram, but it is not a literal path.

The path dependence of work has an important implication for multistep processes such as the one shown in Figure 17.9. The total work done on the gas during the process $1 \rightarrow 2 \rightarrow 3$ must be calculated as $W_{1 \text{ to } 3} = W_{1 \text{ to } 2} + W_{2 \text{ to } 3}$. In this case, $W_{1 \text{ to } 2}$ is negative and $W_{2 \text{ to } 3}$ is positive. Trying to compute the work in a single step, using $\Delta V = V_3 - V_1$, would give you the work of a process that goes directly from 1 to 3. The initial and final states are the same, but the work is *not* the same because work depends on the path followed through the pV diagram.

STOP TO THINK 17.2 Two processes are shown that take an ideal gas from state 1 to state 3. Compare the work done by process A to the work done by process B.

a. $W_A = W_B = 0$
b. $W_A = W_B$ but neither is zero
c. $W_A > W_B$
d. $W_A < W_B$

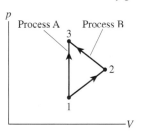

17.3 Heat

Heat is a more elusive concept than work. We use the word "heat" very loosely in the English language, often as synonymous with *hot*. We might say, on a very hot day, that "This heat is oppressive." If your apartment is cold you may say, "Turn up the heat." These expressions date to a time long ago when it was thought that heat was a *substance* with fluid-like properties.

If you place a hot object and a cold object together, they evolve toward a common final temperature. Common sense suggests that "something" flows from the hot object to the cold until equilibrium is achieved. This "heat fluid" was called *caloric*. The notion that objects somehow "contain" heat, and that heat can move around, lingers on in expressions like *heat flow, heat loss,* and *heat capacity.*

One of the first to disagree with this notion, in the late 1700s, was the American-born Benjamin Thompson. Thompson fled to Europe during the American Revolution, settling in Bavaria and later receiving the title Count Rumford. There, while watching the hot metal chips thrown off during the boring of cannons, he began to think about heat. If caloric is a substance, the cannon and borer should eventually run out of caloric and the heat generation ought to decrease with time. But it does not. Rumford noted that the heat generation appears to be "inexhaustible," which is not consistent with the idea of heat as a substance. He concluded that heat is not a substance—it is *motion!*

Rumford was beginning to think along the same lines as had Bernoulli. But Rumford's ideas were speculative and qualitative, hardly a scientific theory, and their implications were not immediately grasped by others. Like Bernoulli's, it would be some time before his insight was recognized and validated.

The turning point was the work of British physicist James Joule in the 1840s. Unlike Bernoulli and Count Rumford, Joule carried out careful experiments to

Heat is the energy transferred in a thermal interaction.

learn how it is that systems change their temperature. Using experiments like those shown in Figure 17.10, Joule found that you can raise the temperature of a beaker of water by two entirely different means:

1. Heating it with a flame, or
2. Doing work on it with a rapidly spinning paddle wheel.

The final state of the water is *exactly* the same in both cases. This implies that heat and work are essentially equivalent to each other. In other words, heat is not a substance. Instead, heat is the *energy* that is transferred from a hotter object to a colder object.

It was Joule who placed the idea of conservation of energy on a firm scientific footing and established it as a law of nature. Heat and work, which previously had been regarded as two completely different phenomena, were seen to be simply two different ways of transferring energy to or from a system. Joule's discoveries vindicated the earlier ideas of Bernoulli and Count Rumford, and they opened the door for rapid advancements in the subject of thermodynamics during the second half of the 19th century.

The flame heats the water.

The spinning paddle does work on the water.

FIGURE 17.10 Joule's experiments to show the equivalence of heat and work.

Thermal Interactions

Heat is the energy transferred between a system and the environment as a consequence of a *temperature difference* between them. Unlike a mechanical interaction in which work is done, heat requires no macroscopic motion of the system. Instead (we'll look at the details in Chapter 18), energy is transferred when the *faster* molecules in the hotter object collide with the *slower* molecules in the cooler object. On average, these collisions cause the faster molecules to lose energy and the slower molecules to gain energy. The net result is that energy is transferred from the hotter object to the colder object. The process itself, whereby energy is transferred between the system and the environment via atomic-level collisions, is called a **thermal interaction.** *Heat* is the energy transferred during a thermal interaction.

When you place a pan of water on the stove, heat is the energy transferred *from* the hotter flame *to* the cooler water. If you place the water in a freezer, heat is the energy transferred from the warmer water to the colder air in the freezer. A system is in **thermal equilibrium** with the environment, or two systems are in thermal equilibrium with each other, if there is no temperature difference.

It is well worthwhile to compare this statement about heat and thermal interactions with the first paragraph about work in Section 17.1. The analogy would be complete if we were able to say that the environment (or an object in the environment) "does heat" on the system. Unfortunately, the English language doesn't work that way. Loosely speaking, we say that the environment "heats" the system.

Like work, **heat is not a state variable.** Heat is not a property of the system. Instead, heat is the amount of energy that moves between the system and the environment during a thermal interaction. It would not be meaningful to talk about a "change of heat." Thus heat appears in the energy equation simply as a value Q, never as ΔQ.

Figure 17.11 shows that Q is positive when energy is transferred *into* the system from the environment. This implies that $T_{env} > T_{sys}$. A negative Q represents heat transfer *from* the system back to the environment when $T_{env} < T_{sys}$. The system is in thermal equilibrium with the environment ($Q = 0$) when $T_{env} = T_{sys}$.

NOTE ▶ For both heat and work, a positive value indicates energy being transferred from the environment to the system. Table 17.1 on the next page summarizes the similarities and differences between work and heat. ◀

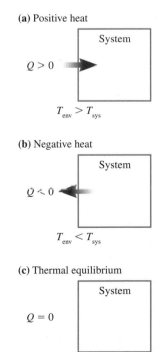

(a) Positive heat

$Q > 0$ — System

$T_{env} > T_{sys}$

(b) Negative heat

$Q < 0$ — System

$T_{env} < T_{sys}$

(c) Thermal equilibrium

$Q = 0$ System

$T_{env} = T_{sys}$

FIGURE 17.11 The sign of heat.

TABLE 17.1 Understanding work and heat

	Work	Heat
Interaction:	Mechanical	Thermal
Requires:	Force and displacement	Temperature difference
Process:	Macroscopic pushes and pulls	Microscopic collisions
Positive value:	$W > 0$ when the force on the system is in the direction of the displacement. Gases are compressed. Energy is transferred in.	$Q > 0$ when the environment is at a higher temperature than the system. Energy is transferred in.
Negative value:	$W < 0$ when the force on the system is opposite the displacement. Gases expand. Energy is transferred out.	$Q < 0$ when the system is at a higher temperature than the environment. Energy is transferred out.
Equilibrium:	A system is in mechanical equilibrium when the environment exerts no net force on it.	A system is in thermal equilibrium when it is at the same temperature as the environment.

Units of Heat

Heat is energy transferred between the system and the environment. Consequently, the SI unit of heat is the joule. Historically, before the connection between heat and work had been recognized, a unit for measuring heat, the calorie, had been defined as

$$1 \text{ calorie} = 1 \text{ cal} = \text{the quantity of heat needed to change}$$
$$\text{the temperature of 1 g of water by 1 °C.}$$

Once Joule established that heat is energy, it was apparent that the calorie is really a unit of energy. Joule used experiments to determine the conversion factor between heat units and more conventional energy units. In today's SI units, that conversion is

$$1 \text{ cal} = 4.186 \text{ J}$$

The calorie you know in relation to food is not the same as the heat calorie. The *food calorie,* abbreviated Cal with a capital C, is

$$1 \text{ food calorie} = 1 \text{ Cal} = 1000 \text{ cal} = 1 \text{ kcal} = 4186 \text{ J}$$

The food calorie measures the food's chemical energy, stored energy that is available for doing work or for keeping your body warm. That extra dessert you ate last night containing 300 Cal has a chemical energy

$$E_{\text{chem}} = 300 \text{ Cal} = 3 \times 10^5 \text{ cal} = 1.26 \times 10^6 \text{ J}$$

We will not use calories in this textbook, but there are some fields of science and engineering where calories are still widely used. All the calculations you learn to do with joules can equally well be done with calories.

The Trouble with Heat

The trouble with heat is twofold: conceptual and linguistic. At the conceptual level, it is important to distinguish between *heat, temperature,* and *thermal energy.* These three ideas are related, but the distinctions between them are crucial. Common language can easily mislead you. If an object slides to a halt because of friction, most people say that the object's kinetic energy is "converted into heat." In fact, heat is not involved in this process. Nowhere was there a transfer of energy due to a temperature difference. Instead, the object's mechanical energy is transformed into the *thermal energy* of the atoms and molecules. In brief,

- Thermal energy is an energy *of the system* due to the motion of its atoms and molecules. It is a *form* of energy. Thermal energy is a state variable, and it makes sense to talk about how E_{th} changes during a process. The system's thermal energy continues to exist even if the system is isolated and not interacting thermally with its environment.
- Heat is energy transferred *between the system* and the environment as they interact. Heat is *not* a particular form of energy, nor is it a state variable. It makes no sense to talk about how heat changes. $Q = 0$ if a system does not interact thermally with its environment. Heat may cause the system's thermal energy to change, but that doesn't mean that heat and thermal energy are the same thing.
- Temperature is a state variable that quantifies the "hotness" or "coldness" of a system. We haven't given a precise definition of temperature, but it is related to the thermal energy *per molecule*. A temperature difference is a requirement for a thermal interaction in which heat energy is transferred between the system and the environment.

It is especially important not to associate an observed temperature increase with heat. Heating a system is one way to change its temperature, but, as Joule showed, not the only way. You can also change the system's temperature by doing work on the system. **Observing the system tells us nothing about the process by which energy enters or leaves the system.**

We have two problems on the linguistic front. One, already alluded to, are those terms such as "heat flow" and "heat capacity" that are vestiges of history. These phrases, which are used even in scientific and technical discourse, incorrectly suggest that heat is a substance that can flow from one object to another or be contained in an object. With experience, scientists and engineers learn to use these phrases without meaning what the phrase, interpreted literally, seems to suggest.

A second problem is that the phrase "to heat" uses the word "heat" as a verb, whereas our definition of "heat" uses the word as a noun. These two uses make no distinction between the energy transferred and the process of transferring the energy. With work, the phase "to *do* work" allows us to separate the process from the energy transferred (i.e., "the work") in the process.

Unfortunately, physics textbooks can't reinvent language. We will try to be very careful in our choice of words and phrases, and will highlight points were the language is potentially confusing or misleading. Being forewarned will help you avoid some of these pitfalls.

STOP TO THINK 17.3   Which one or more of the following processes involves heat?

a. The brakes in your car get hot when you stop.
b. A steel block is held over a candle.
c. You push a rigid cylinder of gas across a frictionless surface.
d. You push a piston into a cylinder of gas, increasing the temperature of the gas.
e. You place a cylinder of gas in hot water. The gas expands, causing a piston to rise and lift a weight. The temperature of the gas does not change.

17.4 The First Law of Thermodynamics

8.8–8.10 Actiⱽ Physics

Heat was the missing piece that we needed to arrive at a completely general statement of the law of conservation of energy. Restating Equation 17.6,

$$\Delta E_{sys} = \Delta E_{mech} + \Delta E_{th} = W + Q$$

Work and heat, two ways of transferring energy between a system and the environment, cause the system's energy to change.

At this point in the text we are not interested in systems that have a macroscopic motion of the system as a whole. Moving macroscopic systems were important to us for many chapters, but now, as we investigate the thermal properties of a system, we would like the system as a whole to rest peacefully on the laboratory bench while we study it. So we will assume, throughout the remainder of Part IV, that $\Delta E_{mech} = 0$.

With this assumption clearly stated, the law of conservation of energy becomes

$$\Delta E_{th} = W + Q \qquad \text{(first law of thermodynamics)} \qquad (17.17)$$

The energy equation, in this form, is called the **first law of thermodynamics** or simply "the first law." The first law is nothing more than a statement about the conservation of energy.

Chapters 10 and 11 introduced the basic energy model. It was called *basic* because it included work but not heat. The first law of thermodynamics has included heat, but it excludes situations where the mechanical energy changes. Figure 17.12 is a pictorial representation of the **thermodynamic energy model** described by the first law. Work and heat are energies transferred between the system and the environment. Energy added to the system (W or Q positive) increases the system's thermal energy ($\Delta E_{th} > 0$). Likewise, the thermal energy decreases when energy is removed from the system.

Two comments are worthwhile:

1. The first law of thermodynamics doesn't tell us anything about the value of E_{th}, only how E_{th} changes. Doing 1 J of work changes the thermal energy by $\Delta E_{th} = 1$ J regardless of whether $E_{th} = 10$ J or 10,000 J.
2. The system's thermal energy isn't the only thing that changes. Work or heat that change the thermal energy also change the pressure, volume, temperature, and other state variables. The first law tells us only about ΔE_{th}. Other laws and relationships must be used to learn how the other state variables change.

The first law is one of the most important analytic tools of thermodynamics. It will be especially important when we look at practical thermodynamics in Chapter 19. We'll use the first law in the remainder of this chapter to study some of the thermal properties of matter.

FIGURE 17.12 The thermodynamic energy model.

Lifting Weights

A simple model can help us understand how the first law of thermodynamics is related to ideal-gas process and pV diagrams. Figure 17.13 shows a gas cylinder with several special properties:

- You can fix the gas volume by inserting the locking pin into the piston. Without the pin, the piston can slide up or down. The piston is massless and insulated.
- You can change the gas pressure by adding or removing masses on top of the piston. Work is done as the piston moves the masses up and down.

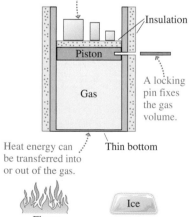

FIGURE 17.13 A gas that can be heated, have work done on it, or be maintained at a constant temperature.

- You can warm or cool the gas by placing the cylinder above a flame or on a block of ice. The thin bottom of the cylinder is the only surface through which heat energy can be transferred.
- All processes are assumed to take place quasi-statically.

When the piston is unlocked, the gas pressure is determined by the atmospheric pressure and by the total mass M on the piston:

$$p_{gas} = p_{atmos} + \frac{Mg}{A} \qquad (17.18)$$

The pressure doesn't change as the piston moves unless you change the mass. (This result is from our discussion of isobaric processes in Chapter 16.) Equation 17.18 is *not* valid when the piston is locked. The pressure with the piston locked could be either higher or lower than Equation 17.18.

EXAMPLE 17.4 An isochoric cooling process

Design and implement a process that will decrease the pressure in the gas cylinder of Figure 17.13 without changing the volume. Describe the steps, then show the process on a pV diagram and as a first-law bar chart.

MODEL To bring about this process:

- Insert the locking pin so that the volume cannot change.
- Place the cylinder on the block of ice. Heat energy will be transferred from the gas to the ice, causing the gas temperature and pressure to fall.
- Remove the cylinder from the ice when the desired pressure is reached.
- Remove masses from the piston until the total mass M balances the new gas pressure. This step must be done before removing the locking pin; otherwise the piston will move when the pin is removed.
- Remove the locking pin.

VISUALIZE This is the isochoric process shown in Figure 17.14a. The final point is on a lower isotherm than the initial point, so $T_f < T_i$. No work is done ($W = 0$) in an isochoric process because the piston doesn't move. Heat energy was transferred out of the gas ($Q < 0$) and the thermal energy of the gas decreased ($\Delta E < 0$) as the temperature fell. From the first law, with $W = 0$, we can write explicitly that $E_{th\,f} = E_{th\,i} + Q$. Figure 17.14b shows this result on a first-law bar chart. We don't know the value of the initial thermal energy $E_{th\,i}$, so the height of the $E_{th\,i}$ bar is arbitrary. But we can show how the thermal energy *changes* as a consequence of the gas process.

In this example, the $E_{th\,f}$ bar shows that the thermal energy has been decreased by the amount of energy that left the system as heat.

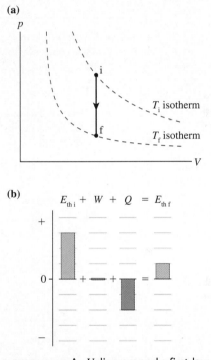

FIGURE 17.14 A pV diagram and a first-law bar chart for a process that decreases the pressure without changing the volume.

EXAMPLE 17.5 An isothermal expansion

Design and implement a process that will increase the volume in the gas cylinder of Figure 17.13 without changing the temperature. Describe the steps, then show the process on a pV diagram and as a first-law bar chart.

MODEL To bring about this process:

- Place the cylinder over the flame. Heat energy will be transferred to the gas and the gas will begin to expand.

- The product pV must remain constant during an isothermal process. Slowly remove masses from the piston to reduce the pressure as the volume increases. The temperature remains constant as heat energy from the flame balances the work done in the expansion.

- Remove the cylinder from the flame when the gas reaches the desired volume.

VISUALIZE This is the isothermal process shown in Figure 17.15a. $\Delta E_{th} = 0$ in an isothermal process ($\Delta T = 0$), so the first law $\Delta E_{th} = W + Q$ can be satisfied only if $W = -Q$. Heat energy is transferred to the gas, but the temperature of the gas doesn't change. Instead, the energy causes the gas to expand and do the work of lifting the masses. The work done *on* the gas by the force holding the piston in place is negative when a gas expands. This information is shown on the first-law bar chart of Figure 17.15b.

ASSESS It is surprising, but true, that we can heat the system without changing its temperature. But to do so, we must have a process in which the energy coming into the system as heat is exactly balanced by energy leaving the system as work.

FIGURE 17.15 A pV diagram and a first-law bar chart for a process that increases the volume without changing the temperature.

STOP TO THINK 17.4   Which first-law bar chart describes the process shown in the pV diagram?

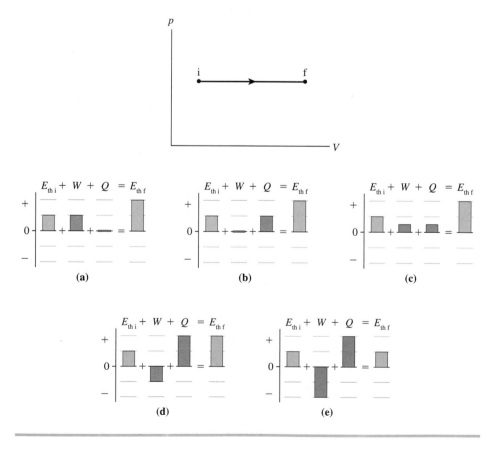

17.5 Thermal Properties of Matter

Joule established that heat and work are energy transferred between a system and its environment. Heat and work are equivalent in the sense that the change of the system is *exactly the same* whether you transfer heat energy to it or do an equal amount of work on it. Adding energy to the system, or removing it, changes the system's thermal energy.

What happens to a system when you change its thermal energy? In this section we'll consider two distinct possibilities:

- The temperature of the system changes.
- The system undergoes a phase change, such as melting or freezing.

Temperature Change and Specific Heat

Suppose you do an experiment in which you add energy to water, either by doing work on it or transferring heat to it. Either way, you will find that adding 4190 J of energy raises the temperature of 1 kg of water by 1 K. If you were fortunate enough to have 1 kg of gold, you need to add only 129 J of energy to raise its temperature by 1 K.

The amount of energy that raises the temperature of 1 kg of a substance by 1 K is called the **specific heat** of that substance. The symbol for specific heat is c. Water has specific heat $c_{water} = 4190$ J/kg K. The specific heat of gold is $c_{gold} = 129$ J/kg K. Specific heat depends only on the material from which an object is made. Table 17.2 provides some specific heats for common liquids and solids. Extensive tables can be found in handbooks of chemical and physical data.

TABLE 17.2 Specific heats and molar specific heats of solids and liquids

Substance	c (J/kg K)	C (J/mol K)
Solids		
Aluminum	900	24.3
Copper	385	24.4
Iron	449	25.1
Gold	129	25.4
Lead	128	26.5
Ice	2090	37.6
Liquids		
Ethyl alcohol	2400	110.4
Mercury	140	28.1
Water	4190	75.4

NOTE ▶ The term *specific heat* does not use the word *heat* in the way that we have defined it. Specific heat is an old idea, dating back to the days of the caloric theory when heat was thought to be a substance contained in the object. The term has continued in use even though our understanding of heat has changed. ◀

If energy c is required to raise the temperature of 1 kg of substance by 1 K, then energy Mc is needed to raise the temperature of mass M by 1 K and $(Mc)\Delta T$ is needed to raise the temperature of mass M by ΔT. In other words, the thermal energy of the system changes by

$$\Delta E_{th} = Mc\Delta T \qquad \text{(temperature change)} \qquad (17.19)$$

when its temperature changes by ΔT. ΔE_{th} can be either positive (thermal energy increases as the temperature goes up) or negative (thermal energy decreases as the temperature goes down). Recall that uppercase M is used for the mass of an entire system while lowercase m is reserved for the mass of an atom or molecule.

NOTE ▶ In practice, ΔT is usually measured in °C. But the Kelvin and the Celsius temperature scales have the same step size, so ΔT in K has exactly the same numerical value as ΔT in °C. Thus

- You do not need to convert temperatures from °C to K if you only need a temperature *change* ΔT.
- You do need to convert anytime you need the actual temperature T. ◀

The first law of thermodynamics, $\Delta E_{th} = W + Q$, allows us to write Equation 17.19 as $Mc\Delta T = W + Q$. In other words, we can change the system's temperature either by heating it or by doing an equivalent amount of work on it. In working with solids and liquids, we almost always change the temperature by heating. If $W = 0$, which we will assume for the rest of this section, then the heat needed to bring about a temperature change ΔT is

$$Q = Mc\Delta T \qquad \text{(temperature change)} \qquad (17.20)$$

Because $\Delta T = \Delta E_{\text{th}}/Mc$, it takes more energy to change the temperature of a substance with a large specific heat than to change the temperature of a substance with a small specific heat. You can think of specific heat as measuring the *thermal inertia* of a substance. Metals, with small specific heats, warm up and cool down quickly. A piece of aluminum foil can be safely held within seconds of removing it from a hot oven. Water, with a very large specific heat, is slow to warm up and slow to cool down. This is fortunate for us. The large thermal inertia of water is essential for the biological processes of life. We wouldn't be here studying physics if water had a small specific heat!

EXAMPLE 17.6 Heating gold

How much heat is required to raise the temperature of 500 g of gold from 10°C to 50°C?

MODEL Heating the gold changes its thermal energy and its temperature.

SOLVE The temperature change is $\Delta T = 40°C = 40$ K. Thus

$$Q = \Delta E_{\text{th}} = Mc\Delta T = (0.500 \text{ kg})(129 \text{ J/kg K})(40 \text{ K})$$

$$= 2580 \text{ J}$$

where we used $c_{\text{gold}} = 129$ J/kg K from Table 17.2. Thus 2580 J of heat is needed.

EXAMPLE 17.7 Quenching hot aluminum in ethyl alcohol

A 50 g aluminum disk at 300°C is placed in 200 cm^3 of ethyl alcohol at 10.0°C, then quickly removed. The aluminum temperature is found to have dropped to 120°C. What is the new temperature of the ethyl alcohol?

MODEL Heat is the energy transferred due to a temperature difference. If we assume that the container holding the alcohol is well insulated, then the disk and the alcohol interact with each other but nothing else. Conservation of energy tells us that the heat energy transferred out of the disk is the heat energy transferred into the alcohol.

SOLVE The temperature change of the disk is $\Delta T_{\text{Al}} = (120°C - 300°C) = -180°C = -180$ K. It is negative because the temperature decreases. The energy removed from the disk is

$$Q_{\text{Al}} = Mc\Delta T = (0.050 \text{ kg})(900 \text{ J/kg K})(-180 \text{ K})$$

$$= -8100 \text{ J}$$

Q_{Al} is negative because the energy is transferred out of the aluminum. The ethyl alcohol *gains* 8100 J of energy, thus $Q_{\text{ethyl}} = +8100$ J. We need to know the mass of the ethyl alcohol. Its density was given in Table 16.1 as $\rho = 790$ kg/m^3, hence its mass is

$$M = \rho V = (790 \text{ kg/m}^3)(200 \times 10^{-6} \text{ m}^3) = 0.158 \text{ kg}$$

The heat from the aluminum causes the alcohol's temperature to change by

$$\Delta T = \frac{Q_{\text{ethyl}}}{Mc} = \frac{8100 \text{ J}}{(0.158 \text{ kg})(2400 \text{ J/kg K})} = 21.4 \text{ K}$$

$$= 21.4°C$$

The ethyl alcohol ends up at temperature

$$T_{\text{f}} = T_{\text{i}} + \Delta T = 10.0°C + 21.4°C = 31.4°C$$

The **molar specific heat** is the amount of energy that raises the temperature of 1 mol of a substance by 1 K. We'll use an uppercase C for the molar specific heat. The heat needed to bring about a temperature change ΔT of n moles of substance is

$$Q = nC\Delta T \qquad (17.21)$$

The molar specific heat C is not simply a conversion of c to other units because the number of moles per kilogram depends on the molar mass of a substance. The number of moles in mass M of a substance is

$$n = \frac{M(\text{in g})}{M_{\text{mol}}} = \frac{1000 \text{ g/kg}}{M_{\text{mol}}(\text{in g/mol})}M(\text{in kg})$$

Thus the molar specific heat is

$$C\,(\text{in J/mol K}) = \frac{M_{\text{mol}}(\text{in g/mol})}{1000\ \text{g/kg}}\,c\,(\text{in J/kg K}) \qquad (17.22)$$

Molar specific heats are shown in Table 17.2. Look at the five elemental solids (excluding ice). All have C very near 25 J/mol K. If we were to expand the table, we would find that most elemental solids have $C \approx 25$ J/mol K. This can't be a coincidence, but what is it telling us? This is a puzzle we will address in Chapter 18, where we will explore thermal energy at the atomic level.

Phase Change and Heat of Transformation

Suppose you start with a system in its solid phase and heat it at a steady rate. Figure 17.16, which you saw in Chapter 16, shows how the system's temperature changes. At first, the temperature increases linearly. This is not hard to understand because Equation 17.20 can be written

$$\text{slope of the } T\text{-versus-}Q \text{ graph} = \frac{\Delta T}{Q} = \frac{1}{Mc} \qquad (17.23)$$

The slope of the graph depends inversely on the system's specific heat. A constant specific heat implies a constant slope and thus a linear graph. In fact, you can measure c from such a graph.

NOTE ▶ The different slopes indicate that the solid, liquid, and gas phases of a substance have different specific heats. ◀

But there are times, shown as horizontal line segments, during which heat is being transferred to the system but the temperature isn't changing. These are *phase changes*. The thermal energy continues to increase during a phase change, but the additional energy goes into breaking molecular bonds rather than speeding up the molecules. **A phase change is characterized by a change in thermal energy without a change in temperature.**

The amount of heat energy that causes 1 kg of a substance to undergo a phase change is called the **heat of transformation** of that substance. For example, laboratory experiments show that 333,000 J of heat are needed to melt 1 kg of ice at 0°C. The symbol for heat of transformation is L. The heat required for the entire system of mass M to undergo a phase change is

$$Q = ML \qquad \text{(phase change)} \qquad (17.24)$$

Heat of transformation is a generic term that refers to any phase change. Two specific heats of transformation are the **heat of fusion** L_{f}, the heat of transformation between a solid and a liquid, and the **heat of vaporization** L_{v}, the heat of transformation between a liquid and a gas. The heat needed for these phase changes is

$$Q = \begin{cases} \pm ML_{\text{f}} & \text{melt/freeze} \\ \pm ML_{\text{v}} & \text{boil/condense} \end{cases} \qquad (17.25)$$

where the $\pm$ indicates that heat must be *added* to the system during melting or boiling but *removed* from the system during freezing or condensing. **You must explicitly include the minus sign when it is needed.**

Table 17.3 on the next page gives the heats of transformation of a few substances. Notice that the heat of vaporization is always much larger than the heat of fusion. We can understand this. Melting breaks just enough molecular bonds to allow the system to lose rigidity and flow. Even so, the molecules in a liquid

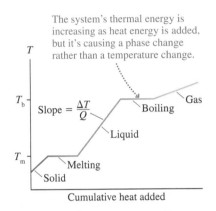

FIGURE 17.16 The temperature of a system that is heated at a steady rate.

Lava—molten rock—undergoes a phase change when it contacts the much colder water. This is one way in which new islands are formed.

remain close together and loosely bonded. Vaporization breaks all bonds completely and sends the molecules flying apart. This process requires a larger increase in the thermal energy and thus a larger quantity of heat.

TABLE 17.3 Melting/boiling temperatures and heats of transformation

Substance	T_m (°C)	L_f (J/kg)	T_b (°C)	L_v (J/kg)
Nitrogen (N_2)	−210	0.26×10^5	−196	1.99×10^5
Ethyl alcohol	−114	1.09×10^5	78	8.79×10^5
Mercury	−39	0.11×10^5	357	2.96×10^5
Water	0	3.33×10^5	100	22.6×10^5
Lead	328	0.25×10^5	1750	8.58×10^5

EXAMPLE 17.8 Turning ice into steam
How much heat is required to change 200 mL of ice at −20°C (a typical freezer temperature) into steam?

MODEL Changing ice to steam requires four steps: Raise the temperature of the ice to 0°C, melt the ice to liquid water at 0°C, raise the water temperature to 100°C, then boil the water to produce steam at 100°C.

SOLVE The mass is $M = \rho V$. The density of ice (from Table 16.1) is 920 kg/m³, and $V = 200$ mL $= 200$ cm³ $= 2.00 \times 10^{-4}$ m³. Thus

$$M = \rho V = (920 \text{ kg/m}^3)(2.00 \times 10^{-4} \text{ m}^3) = 0.184 \text{ kg}$$

The heat needed for each step is

$$Q_1 = Mc_{ice}\Delta T_{ice} = (0.184 \text{ kg})(2090 \text{ J/kg K})(20 \text{ K})$$
$$= 7,700 \text{ J}$$

$$Q_2 = ML_f = (0.184 \text{ kg})(3.33 \times 10^5 \text{ J/kg}) = 61,300 \text{ J}$$

$$Q_3 = Mc_{water}\Delta T_{water} = (0.184 \text{ kg})(4190 \text{ J/kg K})(100 \text{ K})$$
$$= 77,100 \text{ J}$$

$$Q_4 = ML_v = (0.184 \text{ kg})(22.6 \times 10^5 \text{ J/kg}) = 415,800 \text{ J}$$

NOTE ▶ We used the specific heat of ice while warming the system in its solid phase. Then we used the specific heat of *water* while warming the system in its liquid phase. ◀

The total heat required is

$$Q = Q_1 + Q_2 + Q_3 + Q_4 = 562,000 \text{ J}$$

ASSESS Roughly 75% of the heat is used to change the water from a 100°C liquid to a 100°C gas. This is consistent with your experience that it takes much longer for a pan of water to boil away than it does to reach boiling.

STOP TO THINK 17.5   Objects A and B are brought into close thermal contact with each other, but they are well isolated from their surroundings. Initially $T_A = 0$°C and $T_B = 100$°C. The specific heat of A is less than the specific heat of B. The two objects will soon reach a common final temperature T_f. The final temperature is

a. $T_f > 50$°C.
b. $T_f = 50$°C.
c. $T_f < 50$°C.

A
1.0 kg
0°C

B
1.0 kg
100°C

17.6 Calorimetry

At one time or another you've probably put an ice cube into a hot drink to cool it quickly. You were engaged, in a somewhat trial-and-error way, in a practical aspect of heat transfer known as **calorimetry.**

Figure 17.17 shows two systems thermally interacting with each other but isolated from everything else. Suppose they start at different temperatures T_1 and T_2. As you know from experience, heat energy will be transferred from the hotter to the colder system until they reach a common final temperature T_f. The systems will then be in thermal equilibrium and the temperature will not change further.

The insulation prevents any heat energy from being transferred to or from the environment, so energy conservation tells us that any energy leaving the hotter system must enter the colder system. That is, the systems *exchange* energy with no net loss or gain. The concept is straightforward, but to state the idea mathematically we need to be careful with signs.

Let Q_1 be the energy transferred to system 1 as heat. Q_1 is positive if energy *enters* system 1, negative if energy *leaves* system 1. Similarly, Q_2 is the energy transferred to system 2. The fact that the systems are merely exchanging energy can be written $|Q_1| = |Q_2|$. That is, the energy *lost* by the hotter system is the energy *gained* by the colder system. But Q_1 and Q_2 have opposite signs, so $Q_1 = -Q_2$. No energy is exchanged with the environment, hence it makes more sense to write this relationship as

$$Q_{net} = Q_1 + Q_2 = 0 \qquad (17.26)$$

This idea is not limited to the interaction of only two systems. If three or more systems are combined in isolation from the rest of their environment, each at a different initial temperature, they will all come to a common final temperature that can be found from the relationship

$$Q_{net} = Q_1 + Q_2 + Q_3 + \cdots = 0 \qquad (17.27)$$

NOTE ▶ The signs are very important in calorimetry problems. ΔT is always $T_f - T_i$, so ΔT and Q are negative for any system whose temperature decreases. The proper sign of Q for any phase change must be supplied *by you*, depending on the direction of the phase change. ◀

Heat energy is transferred from system 1 to system 2. Energy conservation requires

$$|Q_1| = |Q_2|$$

Opposite signs mean that

$$Q_{net} = Q_1 + Q_2 = 0$$

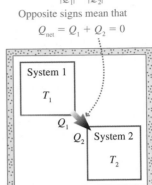

FIGURE 17.17 Two systems interact thermally.

(MP) PROBLEM-SOLVING STRATEGY 17.2 **Calorimetry problems**

MODEL Identify the interacting systems. Assume that they are isolated from the larger environment.

VISUALIZE List known information and identify what you need to find. Convert all quantities to SI units.

SOLVE The mathematical representation, which is a statement of energy conservation, is

$$Q_{net} = Q_1 + Q_2 + \cdots = 0$$

- For systems that undergo a temperature change, $Q = Mc(T_f - T_i)$. Be sure to have the temperatures T_i and T_f in the correct order.
- For systems that undergo a phase change, $Q = \pm ML$. Supply the correct sign by observing whether energy enters or leaves the system during the transition.
- Some systems may undergo a temperature change *and* a phase change. Treat the changes separately. The heat energy is $Q = Q_{\Delta T} + Q_{phase}$.

ASSESS Is the final temperature in the middle? T_f that is higher or lower than all initial temperatures is an indication that something is wrong, usually a sign error.

NOTE ▶ You may have learned to solve calorimetry problems in other courses by writing $Q_{gained} = Q_{lost}$. That is, by balancing heat gained with heat lost. That approach works in simple problems, but it has two drawbacks. First, you often have to "fudge" the signs to make them work. Second, and more serious, you can't extend this approach to a problem with three or more interacting systems. Using $Q_{net} = 0$ is much preferred. ◀

EXAMPLE 17.9 Identifying a metal

200 g of an unknown metal is heated to a temperature of 200°C then dropped into 50 g of water at 20.0°C in an insulated container. The water temperature rises within a few seconds to 39.7°C, then changes no further. Identify the metal.

MODEL Two systems interact thermally. There are temperature changes but no phase changes.

VISUALIZE We know all the initial and final temperatures. For water, $c_w = 4190$ J/kg K is known from Table 17.2. Only the specific heat c_m of the metal is unknown.

SOLVE Energy conservation requires $Q_w + Q_m = 0$. Using $Q = Mc(T_f - T_i)$ for each, we have

$$Q_w + Q_m = M_w c_w (T_f - T_w) + M_m c_m (T_f - T_m) = 0$$

This is easily solved for the unknown specific heat:

$$c_m = \frac{-M_w c_w (T_f - T_w)}{M_m (T_f - T_m)}$$

$$= \frac{-(0.050 \text{ kg})(4190 \text{ J/kg K})(39.7°C - 20°C)}{(0.200 \text{ kg})(39.7°C - 200°C)}$$

$$= 129 \text{ J/kg K}$$

Referring to Table 17.2, we find we have either 200 g of gold or, if we made an ever-so-slight experimental error, 200 g of lead!

NOTE ▶ It's okay to compute ΔT in °C, even though the specific heat is in J/kg K, because a *change* of 1°C is equivalent to a *change* of 1 K. ◀

ASSESS The temperature of the unknown metal changed much more than the temperature of the water. That is, the unknown metal had much less thermal inertia, and thus, as we found, its specific heat should be much less than that of water.

EXAMPLE 17.10 Calorimetry with a phase change

Your 500 mL soda is at 20°C, room temperature, so you add 100 g of ice from the −20°C freezer. Does all the ice melt? If so, what is the final temperature? If not, what fraction of the ice melts? Assume that you have a well-insulated cup.

MODEL We have a thermal interaction between the soda, which is essentially water, and the ice. We need to distinguish between three possible outcomes. If all the ice melts, then $T_f > 0°C$. It's also possible that the soda will cool to 0°C before all the ice has melted, leaving the ice and liquid in equilibrium at 0°C. A third possibility is that the soda will freeze solid before the ice warms up to 0°C. That seems unlikely here, but there are situations, such as the pouring of molten metal out of furnaces, when all the liquid does solidify. We need to distinguish between these before knowing how to proceed.

VISUALIZE All the initial temperatures, masses, and specific heats are known. The final temperature of the combined soda + ice system is unknown.

SOLVE Let's first calculate the heat needed to melt all the ice and leave it as liquid water at 0°C. To do so, we must warm the ice to 0°C, then change it to water. The heat input for this two-stage process is

$$Q_{melt} = M_i c_i (20 \text{ K}) + M_i L_f = 37,500 \text{ J}$$

where L_f is the heat of fusion of water. It is used as a *positive* quantity because we must *add* heat to melt the ice. Next, let's calculate how much heat energy will leave the soda if it cools

all the way to 0°C. The volume is $V = 500$ mL $= 5 \times 10^{-4}$ m^3 and thus the mass is $M_s = \rho V = 0.500$ kg. The heat is

$$Q_{cool} = M_s c_w (-20 \text{ K}) = -41,900 \text{ J}$$

where $\Delta T = -20$ K because the temperature decreases. Because $|Q_{cool}| > Q_{melt}$, the soda has sufficient energy to melt all the ice. Hence the final state will be all liquid at $T_f > 0$. (Had we found $|Q_{cool}| < Q_{melt}$, then the final state would have been an ice-liquid mixture at 0°C.)

Energy conservation requires $Q_{ice} + Q_{soda} = 0$. The heat Q_{ice} consists of three terms: warming the ice to 0°C, melting the ice to water at 0°C, then warming the 0°C water to T_f. The mass will still be M_i in the last of these steps, because it is the "ice system," but we need to use the specific heat of *liquid water*. Thus

$$Q_{ice} + Q_{soda} = \left[M_i c_i (20 \text{ K}) + M_i L_f + M_i c_w (T_f - 0°C) \right]$$
$$+ M_s c_w (T_f - 20°C) = 0$$

We've already done part of the calculation, allowing us to write

$$37,500 \text{ J} + M_i c_w (T_f - 0°C) + M_s c_w (T_f - 20°C) = 0$$

Solving for T_f gives

$$T_f = \frac{20 M_s c_w - 37,500}{M_i c_w + M_s c_w} = 1.7°C$$

ASSESS As expected, the soda has been cooled to nearly the freezing point.

EXAMPLE 17.11 Three interacting systems

A 200 g piece of iron at 120°C and a 150 g piece of copper at −50°C are dropped into an insulated beaker containing 300 g of ethyl alcohol at 20°C. What is the final temperature?

MODEL Here you can't use a simple $Q_{\text{gained}} = Q_{\text{lost}}$ approach because you don't know whether the alcohol is going to warm up or cool down.

VISUALIZE All the initial temperatures, masses, and specific heats are known. We need to find the final temperature.

SOLVE Energy conservation requires

$$Q_i + Q_c + Q_e = M_i c_i (T_f - 120°C) + M_c c_c (T_f - (-50°C))$$
$$+ M_e c_e (T_f - 20°C) = 0$$

Solving for T_f gives

$$T_f = \frac{120 M_i c_i - 50 M_c c_c + 20 M_e c_e}{M_i c_i + M_c c_c + M_e c_e} = 25.7°C$$

ASSESS The temperature is between the initial iron and copper temperatures, as expected. It turns out that the alcohol warms up ($Q_e > 0$), but we had no way to know this without doing the calculation.

17.7 The Specific Heats of Gases

Specific heats are given in Table 17.2 for solids and liquids. Gases are harder to characterize because the heat required to cause a specified temperature change depends on the *process* by which the gas changes state.

Figure 17.18 shows two isotherms on the pV diagram for a gas. Processes A and B, which start on the T_i isotherm and end on the T_f isotherm, have the *same* temperature change $\Delta T = T_f - T_i$. But process A, which takes place at constant volume, requires a *different* amount of heat than does process B, which occurs at constant pressure. The reason is that work is done in process B but not in process A. This is a situation that we are now equipped to analyze.

It is useful to define two different versions of the specific heat of gases, one for constant volume (isochoric) processes and one for constant pressure (isobaric) processes. We will define these as molar specific heats because we usually do gas calculations using moles instead of mass. The quantity of heat needed to change the temperature of n moles of gas by ΔT is

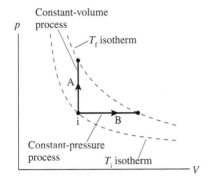

Constant-volume process

T_f isotherm

Constant-pressure process

T_i isotherm

FIGURE 17.18 Processes A and B have the same ΔT and the same ΔE_{th}, but they require different amounts of heat.

$$Q = nC_V \Delta T \quad \text{(temperature change at constant volume)}$$
$$Q = nC_P \Delta T \quad \text{(temperature change at constant pressure)}$$

(17.28)

where C_V is the **molar specific heat at constant volume** and C_P is the **molar specific heat at constant pressure**. Table 17.4 gives the values of C_V and C_P for a few common monatomic and diatomic gases. The units are J/mol K.

NOTE ▶ Equation 17.28 applies to two specific ideal-gas processes. In a general gas process, for which neither p nor V are constant, we have no direct way to relate Q to ΔT. In that case, the heat must be found indirectly from the first law as $Q = \Delta E_{\text{th}} - W$. ◀

TABLE 17.4 Molar specific heats of gases (J/mol K)

Gas	C_P	C_V	$C_P - C_V$
Monatomic Gases			
He	20.8	12.5	8.3
Ne	20.8	12.5	8.3
Ar	20.8	12.5	8.3
Diatomic Gases			
H_2	28.7	20.4	8.3
N_2	29.1	20.8	8.3
O_2	29.2	20.9	8.3

EXAMPLE 17.12 Heating and cooling a gas

Three moles of O_2 gas are at 20.0°C. 600 J of heat energy are transferred to the gas at constant pressure, then 600 J are removed at constant volume. What is the final temperature? Show the process on a pV diagram.

MODEL O_2 is a diatomic ideal gas. The gas is heated as an isobaric process, then cooled as an isochoric process.

SOLVE The heat transferred during the constant-pressure process causes a temperature rise:

$$\Delta T = T_2 - T_1 = \frac{Q}{nC_P} = \frac{600 \text{ J}}{(3.0 \text{ mol})(29.2 \text{ J/mol K})} = 6.8°C$$

where C_P for oxygen was taken from Table 17.4. Heating leaves the gas at temperature $T_2 = T_1 + \Delta T = 26.8°C$. The temperature then falls as heat is removed during the constant-volume process:

$$\Delta T = T_3 - T_2 = \frac{Q}{nC_V} = \frac{(-600\text{ J})}{(3.0\text{ mol})(20.9\text{ J/mol K})} = -9.5°C$$

We used a *negative* value for Q because heat energy is transferred from the gas to the environment. The final temperature of the gas is $T_3 = T_2 + \Delta T = 17.3°C$. Figure 17.19 shows the process on a pV diagram. The gas expands (moves horizontally on the diagram) as heat is added, then cools at constant volume (moves vertically on the diagram) as heat is removed.

ASSESS The final temperature is lower than the initial temperature because $C_P > C_V$.

FIGURE 17.19 The pV diagram for Example 17.12.

EXAMPLE 17.13 Calorimetry with a gas and a solid

The interior volume of a 200 g hollow aluminum box is 800 cm³. The box contains nitrogen gas at STP. A 20 cm³ block of copper at a temperature of 300°C is placed inside the box, then the box is sealed. What is the final temperature?

MODEL This example has three interacting systems: the aluminum box, the nitrogen gas, and the copper block. They must all come to a common final temperature T_f.

VISUALIZE The box and gas have the same initial temperature: $T_{Al} = T_{N2} = 0°C$. The box doesn't change size, so this is a constant-volume process for the gas. The final temperature is unknown.

SOLVE Although one of the systems is now a gas, the calorimetry equation $Q_{net} = Q_{Al} + Q_{N2} + Q_{Cu} = 0$ is still appropriate. In this case,

$$Q_{net} = m_{Al}c_{Al}(T_f - T_{Al}) + n_{N2}C_V(T_f - T_{N2})$$
$$+ m_{Cu}c_{Cu}(T_f - T_{Cu}) = 0$$

Notice that we used masses and specific heats for the solids but moles and the molar specific heat for the gas. We used C_V because this is a constant-volume process. Solving for T_f gives

$$T_f = \frac{m_{Al}c_{Al}T_{Al} + n_{N2}C_VT_{N2} + m_{Cu}c_{Cu}T_{Cu}}{m_{Al}c_{Al} + n_{N2}C_V + m_{Cu}c_{Cu}}$$

The specific heat values are found in Tables 17.2 and 17.4. The mass of the copper is

$$m_{Cu} = \rho_{Cu}V_{Cu} = (8920\text{ kg/cm}^3)(20 \times 10^{-6}\text{ m}^3) = 0.178\text{ kg}$$

The number of moles of the gas is found from the ideal-gas law, using the initial conditions. Notice that inserting the copper block *displaces* 20 cm³ of gas, hence the gas volume is only $V = 780\text{ cm}^3 = 7.80 \times 10^{-4}\text{ m}^3$. Thus

$$n_{N2} = \frac{pV}{RT} = 0.0348\text{ mol}$$

Computing the final temperature from the above equation gives $T_f = 83°C$.

C_P and C_V

8.7 **Act'v Physics**

You may have noticed two curious features in Table 17.4. First, the molar specific heats of monatomic gases are *all alike*. And the molar specific heats of diatomic gases, while different from monatomic gases, are again *very nearly alike*. We saw a similar feature in Table 17.2 for the molar specific heats of solids. Second, the *difference* $C_P - C_V = 8.3\text{ J/mol K}$ is the same in every case. And, most puzzling of all, the value of $C_P - C_V$ appears to be equal to the universal gas constant R! Why should this be?

The relationship between C_V and C_P hinges on one crucial idea: ΔE_{th}, **the change in the thermal energy of a gas, is the same for** *any* **two processes that have the same** ΔT. The thermal energy of a gas is associated with temperature, so any process that changes the gas temperature from T_i to T_f has the same ΔE_{th} as any other process that goes from T_i to T_f. Furthermore, the first law $\Delta E_{th} = Q + W$ tells us that a gas cannot distinguish between heat and work. The system's thermal energy changes in response to energy added to or removed from the system, but the response of the gas is the same whether you heat the system, do work on the system, or do some combination of both. Thus **any two processes that change the thermal energy of the gas by ΔE_{th} will cause the same temperature change ΔT.**

With that in mind, look back at Figure 17.18. Both gas processes have the same ΔT, so both have the same value of ΔE_{th}. Process A is an isochoric process in which no work is done (the piston doesn't move), so the first law for this process is

$$(\Delta E_{th})_A = W + Q = 0 + Q_{const\ vol} = nC_V\Delta T \qquad (17.29)$$

Process B is an isobaric process. You learned earlier that the work done on the gas during an isobaric process is $W = -p\Delta V$. Thus

$$(\Delta E_{th})_B = W + Q = -p\Delta V + Q_{const\ press} = -p\Delta V + nC_P\Delta T \quad (17.30)$$

$(\Delta E_{th})_B = (\Delta E_{th})_A$, because both have the same ΔT, so we can equate the right sides of Equations 17.29 and 17.30:

$$-p\Delta V + nC_P\Delta T = nC_V\Delta T \qquad (17.31)$$

For the final step, we can use the ideal-gas law $pV = nRT$ to relate ΔV and ΔT during process B. For any gas process,

$$\Delta(pV) = \Delta(nRT) \qquad (17.32)$$

For a constant-pressure process, where p is constant, Equation 17.32 becomes

$$p\Delta V = nR\Delta T \qquad (17.33)$$

Substituting this expression for $p\Delta V$ into Equation 17.31 gives

$$-nR\Delta T + nC_P\Delta T = nC_V\Delta T \qquad (17.34)$$

The $n\Delta T$ cancels, and we are left with

$$C_P = C_V + R \qquad (17.35)$$

This result, which applies to all ideal gases, is exactly what we see in the data of Table 17.4.

But that's not the only conclusion we can draw. Equation 17.29 found that $\Delta E_{th} = nC_V\Delta T$ for a constant-volume process. But we had just noted that ΔE_{th} is the same for *all* gas processes that have the same ΔT. Consequently, this expression for ΔE_{th} is equally true for any other process. That is

$$\Delta E_{th} = nC_V\Delta T \qquad \text{(any ideal-gas process)} \qquad (17.36)$$

Compare this result to Equation 17.28. We first made a distinction between constant volume and constant-pressure processes, but now we're saying that Equation 17.36 is true for any process. Are we contradicting ourselves? No, the difference lies in what you need to calculate.

- The change in thermal energy when the temperature changes by ΔT is the same for any process. That's Equation 17.36.
- The *heat* required to bring about the temperature change depends on what the process is. That's Equation 17.28. An isobaric process requires more heat than an isochoric process that produces the same ΔT.

The reason for the difference is seen by writing the first law as $Q = \Delta E_{th} - W$. In an isochoric process, where $W = 0$, *all* the heat input is used to increase the gas temperature. But in an isobaric process, some of the energy that enters the system as heat then leaves the system as work ($W < 0$) done by the expanding gas. Thus more heat is needed to produce the same ΔT.

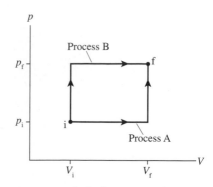

FIGURE 17.20 Is the heat input along these two paths the same or different?

FIGURE 17.21 The relationship of three important processes to the first law of thermodynamics.

Heat Depends on the Path

Consider the two ideal-gas processes shown in Figure 17.20. Even though the initial and final states are the same, the heat added during these two processes is *not* the same. We can use the first law $\Delta E_{th} = W + Q$ to see why.

The thermal energy is a state variable. That is, its value depends on the state of the gas, not the process by which the gas arrived at that state. Thus $\Delta E_{th} = E_{th\,f} - E_{th\,i}$ is the same for both processes. If ΔE_{th} is the same for processes A and B, then $W_A + Q_A = W_B + Q_B$.

You learned in Section 17.2 that the work done during an ideal-gas process depends on the path in the pV diagram. There's more area under the process B curve, so $|W_B| > |W_A|$. Both values of W are negative, because the gas expands, so W_B is more negative than W_A. Consequently, $W_A + Q_A$ can equal $W_B + Q_B$ only if $Q_B > Q_A$. **The heat added or removed during an ideal-gas process depends on the path followed through the pV diagram.**

Adiabatic Processes

We can conclude this section by introducing a fourth basic ideal-gas process. Of the three processes we've looked at thus far—isochoric, isobaric, and isothermal—an isochoric process does no work ($W = 0$) and the isothermal process has no change in thermal energy ($\Delta E_{th} = 0$). Work and thermal energy are two of the three quantities that appear in the first law of thermodynamics. This quite naturally leads us to wonder if there is a process for which no heat is transferred. There is.

An ideal-gas process in which no heat energy is transferred between the gas and the environment ($Q = 0$) is called an **adiabatic process.** The word comes from a Greek root meaning "heat does not pass through." Figure 17.21 compares an adiabatic process to isothermal and isochoric processes. In practice, there are two ways that an adiabatic process can come about. First, a gas cylinder can be completely surrounded by thermal insulation, such as thick pieces of Styrofoam. The environment can interact mechanically with the gas by pushing or pulling on the insulated piston, but there is no thermal interaction.

Second, the gas can be expanded or compressed very rapidly in what we call an *adiabatic expansion* or an *adiabatic compression*. In a rapid process there is essentially no time for heat to be transferred between the gas and the environment. We've already alluded to the idea that heat is transferred via atomic-level collisions. These collisions take time. If you stick one end of a copper rod into a flame, the other end will eventually get too hot to hold—but not instantly. Some amount of time—many seconds in this example—is required for heat to be transferred from one end to the other. A process that takes place faster than the heat can be transferred is adiabatic.

> NOTE ▶ You may recall reading in Chapter 16 that we are going to study only quasi-static processes, processes that proceed slowly enough to remain essentially in equilibrium at all times. Now we're proposing to study processes that take place very rapidly. Isn't this a contradiction? Yes, to some extent it is. What we need to establish are the appropriate time scales. How slow must a process go to be quasi-static? How fast must it go to be adiabatic? These types of calculations must be deferred to a more advanced course. It turns out—fortunately!—that many practical applications, such as the compression strokes in gasoline and diesel engines, are fast enough to be adiabatic yet slow enough to still be considered quasi-static. ◀

For an adiabatic process, with $Q = 0$, the first law of thermodynamics is $\Delta E_{th} = W$. Compressing a gas adiabatically ($W > 0$) increases the thermal energy. Thus **an adiabatic compression raises the temperature of a gas.** A gas that expands adiabatically ($W < 0$) gets colder as its thermal energy decreases.

Thus **an adiabatic expansion lowers the temperature of a gas.** You can use an adiabatic process to change the gas temperature without using heat!

 8.6, 8.11

The work done on a gas in an adiabatic process goes entirely to changing the thermal energy of the gas. But we just found that $\Delta E_{th} = nC_V\Delta T$ for *any* process. Thus

$$W = nC_V\Delta T \qquad \text{(adiabatic process)} \qquad (17.37)$$

Equation 17.37 joins with the equations we derived earlier for the work done in isochoric, isobaric, and isothermal processes.

Gas processes can be represented as trajectories in the pV diagram. For example, a gas moves along a hyperbola during an isothermal process. How does an adiabatic process appear in a pV diagram? The result is more important than the derivation, which is a bit tedious, so we'll begin with the answer and then, at the end of this section, show where it comes from.

First, define the **specific heat ratio** γ (lowercase Greek gamma) to be

$$\gamma = \frac{C_P}{C_V} = \begin{cases} 1.67 & \text{monatomic gas} \\ 1.40 & \text{diatomic gas} \end{cases} \qquad (17.38)$$

The specific heat ratio has many uses in thermodynamics. The values of γ for monatomic and diatomic gases are simple and independent of the gas. Values of γ for more complex gases, such as steam or gasoline vapor, are tabulated in books on engineering thermodynamics. Notice that γ is dimensionless.

An adiabatic process is one in which

$$pV^\gamma = \text{constant} \qquad \text{or} \qquad p_f V_f^\gamma = p_i V_i^\gamma \qquad (17.39)$$

This is similar to the isothermal $pV = \text{constant}$, but somewhat more complex due to the exponent γ.

The curves found by graphing $p = \text{constant}/V^\gamma$ are called **adiabats.** In Figure 17.22 you see that the two adiabats are steeper than the hyperbolic isotherms. An adiabatic process moves along an adiabat in the same way that an isothermal process moves along an isotherm. You can see that the temperature falls during an adiabatic expansion and rises during an adiabatic compression.

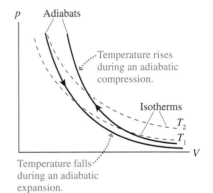

FIGURE 17.22 An adiabatic process moves along pV curves called *adiabats.*

EXAMPLE 17.14 An adiabatic compression

Air containing gasoline vapor is admitted into the cylinder of an internal combustion engine at 1.0 atm pressure and 30°C. The piston rapidly compresses the gas from 500 cm³ to 50 cm³, a *compression ratio* of 10.

a. What are the final temperature and pressure of the gas?
b. Show the compression process on a pV diagram.
c. How much work is done to compress the gas?

MODEL The compression is rapid, with insufficient time for heat to be transferred from the gas to the environment, so we will model it as an adiabatic compression. We'll treat the gas as if it were 100% air.

SOLVE

a. We know the initial pressure and volume, and we know the volume after the compression. For an adiabatic process, where pV^γ remains constant, the final pressure is

$$p_f = p_i\left(\frac{V_i}{V_f}\right)^\gamma = (1.0 \text{ atm})(10)^{1.40} = 25.1 \text{ atm}$$

Air is a mixture of N_2 and O_2, diatomic gases, so we used $\gamma = 1.40$. We can now find the temperature by using the ideal-gas law:

$$T_f = T_i \frac{p_f \, V_f}{p_i \, V_i} = (303 \text{ K})(25.1)\left(\frac{1}{10}\right) = 761 \text{ K} = 488°C$$

Temperature *must* be in kelvins for doing gas calculations such as these.

b. Figure 17.23 shows the pV diagram. The 30°C and 488°C isotherms are included to show how the temperature changes during the process.

FIGURE 17.23 The adiabatic compression of the gas in an internal combustion engine.

c. The work done is $W = nC_V\Delta T$, with $\Delta T = 458$ K. The number of moles is found from the ideal-gas law and the initial conditions:

$$n = \frac{p_i V_i}{RT_i} = 0.0201 \text{ mol}$$

Thus the work done to compress the gas is

$$W = nC_V\Delta T = (0.0201 \text{ mol})(20.8 \text{ J/mol K})(458 \text{ K}) = 192 \text{ J}$$

ASSESS The temperature rises dramatically during the compression stroke of an engine. But the higher temperature has nothing to do with heat! **The temperature and thermal energy of the gas are increased not by heating the gas but by doing work on it.** This is an important idea to understand.

If we use the ideal-gas-law expression $p = nRT/V$ in the adiabatic equation $pV^\gamma = $ constant, we see that $TV^{\gamma-1}$ is also constant during an adiabatic process. Thus another useful equation for adiabatic processes is

$$T_f V_f^{\gamma-1} = T_i V_i^{\gamma-1} \tag{17.40}$$

Proof of Equation 17.39

Now let's see where Equation 17.39 comes from. Consider an adiabatic process in which an infinitesimal amount of work dW done on a gas causes an infinitesimal change in the thermal energy. For an adiabatic process, with $dQ = 0$, the first law of thermodynamics is

$$dE_{th} = dW \tag{17.41}$$

We can use Equation 17.36, which is valid for *any* gas process, to write $dE_{th} = nC_V dT$. Earlier in the chapter we found that the work done during a small volume change is $dW = -p\,dV$. With these substitutions, Equation 17.41 becomes

$$nC_V dT = -p\,dV \tag{17.42}$$

The ideal-gas law can now be used to write $p = nRT/V$. The n cancels, and the C_V can be moved to the other side of the equation to give

$$\frac{dT}{T} = -\frac{R}{C_V}\frac{dV}{V}$$ (17.43)

We're going to integrate Equation 17.43, but anticipating the need for $\gamma = C_P/C_V$ we can first use the fact that $C_P = C_V + R$ to write

$$\frac{R}{C_V} = \frac{C_P - C_V}{C_V} = \frac{C_P}{C_V} - 1 = \gamma - 1$$ (17.44)

Now integrate Equation 17.43 from the initial state i to the final state f:

$$\int_{T_i}^{T_f}\frac{dT}{T} = -(\gamma - 1)\int_{V_i}^{V_f}\frac{dV}{V}$$ (17.45)

Carrying out the integration gives

$$\ln\left(\frac{T_f}{T_i}\right) = \ln\left(\frac{V_i}{V_f}\right)^{\gamma-1}$$ (17.46)

where we used the logarithm properties $\log a - \log b = \log(a/b)$ and $c\log a = \log(a^c)$.

Taking the exponential of both sides now gives

$$\left(\frac{T_f}{T_i}\right) = \left(\frac{V_i}{V_f}\right)^{\gamma-1}$$ (17.47)

$$T_f V_f^{\gamma-1} = T_i V_i^{\gamma-1}$$

This was Equation 17.40. Writing $T = pV/nR$ and canceling $1/nR$ from both sides of the equation gives Equation 17.39:

$$p_f V_f^{\gamma} = p_i V_i^{\gamma}$$ (17.48)

This was a lengthy derivation, but it is good practice at seeing how the ideal-gas law and the first law of thermodynamics can work together to yield results of great importance.

STOP TO THINK 17.6   For the two processes shown, which of the following is true:

a. $Q_A > Q_B$.
b. $Q_A = Q_B$.
c. $Q_A < Q_B$.

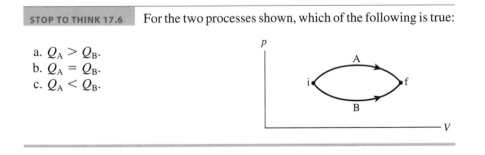

SUMMARY

The goal of Chapter 17 has been to expand our understanding of energy and to develop the first law of thermodynamics as a general statement of energy conservation.

GENERAL PRINCIPLES

First Law of Thermodynamics

$$\Delta E_{th} = W + Q$$

The first law is a general statement of energy conservation.

Work W and heat Q depend on the process by which the system is changed.

The change in the system depends only on the total energy exchanged $W + Q$, not on the process.

Energy

Thermal energy E_{th} Microscopic energy of moving molecules and stretched molecular bonds. ΔE_{th} depends on the initial/final states but is independent of the process.

Work W Energy transferred to the system by forces in a mechanical interaction.

Heat Q Energy transferred to the system via atomic-level collisions when there is a temperature difference. A thermal interaction.

IMPORTANT CONCEPTS

The work done on a gas is

$$W = -\int_{V_i}^{V_f} p\,dV$$

$$= -(\text{area under the } pV \text{ curve})$$

An adiabatic process is one for which $Q = 0$. Gases move along an **adiabat** for which $pV^\gamma = \text{constant}$, where $\gamma = C_P/C_V$ is the **specific heat ratio.** An adiabatic process changes the temperature of the gas without heating it.

Calorimetry When two or more systems interact thermally, they come to a common final temperature determined by

$$Q_{net} = Q_1 + Q_2 + \ldots = 0$$

The **heat of transformation** L is the energy needed to cause 1 kg of substance to undergo a phase change

$$Q = \pm ML$$

The **specific heat** c of a substance is the energy needed to raise the temperature of 1 kg by 1 K.

$$Q = Mc\Delta T$$

The **molar specific heat** C is the energy needed to raise the temperature of 1 mol by 1 K.

$$Q = nC\Delta T$$

The molar specific heat of gases depends on the *process* by which the temperature is changed:

$C_V = $ molar specific heat at **constant volume.**

$C_P = $ molar specific heat at **constant pressure.**

$C_P = C_V + R$, where R is the universal gas constant.

SUMMARY OF BASIC GAS PROCESSES

Process	Definition	Stays constant	Work	Heat
Isochoric	$\Delta V = 0$	V and p/T	$W = 0$	$Q = nC_V\Delta T$
Isobaric	$\Delta p = 0$	p and V/T	$W = -p\Delta V$	$Q = nC_P\Delta T$
Isothermal	$\Delta T = 0$	T and pV	$W = -nRT\ln(V_f/V_i)$	$\Delta E_{th} = 0$
Adiabatic	$Q = 0$	pV^γ	$W = \Delta E_{th}$	$Q = 0$
All gas processes		Ideal-gas law	$pV = nRT$	
		First law	$\Delta E_{th} = W + Q = nC_V\Delta T$	

TERMS AND NOTATION

internal energy, E_{int}	first law of thermodynamics	calorimetry
work, W	thermodynamic energy model	molar specific heat at constant volume, C_V
mechanical interaction	specific heat, c	molar specific heat at constant pressure, C_P
mechanical equilibrium	molar specific heat, C	adiabatic process
heat, Q	heat of transformation, L	specific heat ratio, γ
thermal interaction	heat of fusion, L_f	adiabat
thermal equilibrium	heat of vaporization, L_v	

EXERCISES AND PROBLEMS

Exercises

Section 17.1 It's All About Energy

Section 17.2 Work in Ideal-Gas Processes

1. The average speed of the atoms in a 2.0 g sample of helium gas is 700 m/s. Estimate the thermal energy of the sample.
2. An 8.0 g sample of oxygen gas has 1700 J of thermal energy. Estimate the average speed of an oxygen molecule.
3. How much work is done on the gas in the process shown in Figure Ex17.3?

FIGURE EX17.3

4. How much work is done on the gas in the process shown in Figure Ex17.4?

FIGURE EX17.4

5. 80 J of work are done on the gas in the process shown in Figure Ex17.5. What is V_1 in cm³?

FIGURE EX17.5

6. A 2000 cm³ container holds 0.10 mol of helium gas at 300°C. How much work must be done to compress the gas to 1000 cm³ at (a) constant pressure and (b) constant temperature? (c) Show and label both processes on a single pV diagram.

Section 17.3 Heat

Section 17.4 The First Law of Thermodynamics

7. Draw a first-law bar chart (see Figure 17.14) for the gas process in Figure Ex17.7.

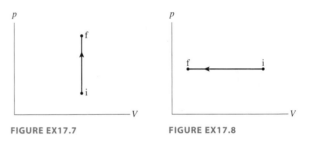

FIGURE EX17.7   **FIGURE EX17.8**

8. Draw a first-law bar chart (see Figure 17.14) for the gas process in Figure Ex17.8.
9. Draw a first-law bar chart (see Figure 17.14) for the gas process in Figure Ex17.9.

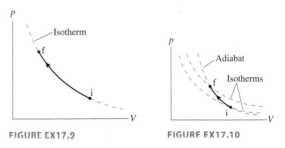

FIGURE EX17.9   **FIGURE EX17.10**

10. Draw a first-law bar chart (see Figure 17.14) for the gas process in Figure Ex17.10.
11. Design and implement a process that will use the gas cylinder of Figure 17.13 to carry out the process shown in Figure Ex17.11. Describe the steps, then show the process as a first-law bar chart.

FIGURE EX17.11

12. Design and implement a process that will use the gas cylinder of Figure 17.13 to carry out the process shown in Figure Ex17.12. Describe the steps, then show the process as a first-law bar chart.

FIGURE EX17.12

13. 500 J of work are done on a system in a process that decreases the system's thermal energy by 200 J. How much heat energy is transferred to or from the system?

14. A gas is compressed from 600 cm³ to 200 cm³ at a constant pressure of 400 kPa. At the same time, 100 J of heat energy is transferred out of the gas. What is the change in thermal energy of the gas during this process?

Section 17.5 Thermal Properties of Matter

15. How much energy must be removed from a 6.0 cm × 6.0 cm × 6.0 cm block of ice to cool it from 0°C to −30°C?

16. A rapidly spinning paddle wheel raises the temperature of 200 mL of water from 21°C to 25°C. How much (a) work is done and (b) heat is transferred in this process?

17. How much heat is needed to change 20 g of mercury at 20°C into mercury vapor at the boiling point?

18. a. 100 J of heat energy are transferred to 20 g of mercury. By how much does the temperature increase?
    b. How much heat is needed to raise the temperature of 20 g of water by the same amount?

19. A beaker contains 200 mL of ethyl alcohol at 20°C. What is the minimum amount of energy that must be removed to produce solid ethyl alcohol?

20. What is the maximum mass of lead you could melt with 1000 J of heat, starting from 20°C?

Section 17.6 Calorimetry

21. 30 g of copper pellets are removed from a 300°C oven and immediately dropped into 100 mL of water at 20°C in an insulated cup. What will the new water temperature be?

22. A copper block is removed from a 300°C oven and dropped into 1.0 L of water at 20.0°C. The water quickly reaches 25.5°C and then remains at that temperature. What is the mass of the copper block?

23. A 50.0 g thermometer is used to measure the temperature of 200 mL of water. The specific heat of the thermometer, which is mostly glass, is 750 J/kg K, and it reads 20.0°C while lying on the table. After being completely immersed in the water, the thermometer's reading stabilizes at 71.2°C. What was the actual water temperature before it was measured?

24. A 750 g aluminum pan is removed from the stove and plunged into a sink filled with 10 L of water at 20.0°C. The water temperature quickly rises to 24.0°C. What was the initial temperature of the pan in °C and in °F?

25. A 500 g metal sphere is heated to 300°C, then dropped into a beaker containing 300 cm³ of mercury at 20.0°C. A short time later the mercury temperature stabilizes at 99.0°C. Identify the metal.

Section 17.7 The Specific Heats of Gases

26. A container holds 1.0 g of argon at a pressure of 8.0 atm.
    a. How much heat is required to increase the temperature by 100°C at constant volume?
    b. How much will the temperature increase if this amount of heat energy is transferred to the gas at constant pressure?

27. A container holds 1.0 g of oxygen at a pressure of 8.0 atm.
    a. How much heat is required to increase the temperature by 100°C at constant pressure?
    b. How much will the temperature increase if this amount of heat energy is transferred to the gas at constant volume?

28. The temperature of 2.0 g of helium is increased at constant volume by ΔT. What mass of oxygen can have its temperature increased by the same amount at constant volume using the same amount of heat?

29. A gas cylinder holds 0.10 mol of O_2 at 150°C and a pressure of 3.0 atm. The gas expands adiabatically until the volume is doubled. What are the final (a) pressure and (b) temperature?

30. A gas cylinder holds 0.10 mol of O_2 at 150°C and a pressure of 3.0 atm. The gas expands adiabatically until the pressure is halved. What are the final (a) volume and (b) temperature?

31. What compression ratios V_i/V_f will raise the temperature of (a) air and (b) argon from 30°C to 850°C in an adiabatic process?

Problems

32. A 1530 cm³ container holds 100°C steam at a pressure of 2.0 atm. How much heat is needed to change this water vapor into ice at −20°C?

33. A 5.0-m-diameter garden pond is 30 cm deep. Solar energy is incident on the pond at an average rate of 400 W/m². If the water absorbs all the solar energy and does not exchange energy with its surroundings, how many hours will it take to warm from 15°C to 25°C?

34. An 11 kg bowling ball at 0°C is dropped into a tub containing a mixture of ice and water. A short time later, when a new equilibrium has been established, there are 5.0 g less ice. From what height was the ball dropped? Assume no water or ice splashes out.

35. The burner on an electric stove has a power output of 2.0 kW. A 750 g stainless steel teakettle is filled with 20°C water and placed on the already hot burner. If it takes 3.0 min for the water to reach a boil, what volume of water, in cm³, was in the kettle? Stainless steel is mostly iron, so you can assume its specific heat is that of iron.

36. Two cars collide head on while each is traveling at 80 km/hr. Suppose all their kinetic energy is transformed into the thermal energy of the wrecks. What is the temperature increase of each car? You can assume that each car's specific heat is that of iron.

37. 10 g of aluminum at 200°C and 20 g of copper are dropped into 50 cm³ of ethyl alcohol at 15°C. The temperature quickly comes to 25°C. What was the initial temperature of the copper?

38. A 100 g ice cube at −10°C is placed in an aluminum cup whose initial temperature is 70°C. The system comes to an equilibrium temperature of 20°C. What is the mass of the cup?

39. 512 g of an unknown metal at a temperature of 15°C is dropped into a 100 g aluminum container holding 325 g of water at 98°C. A short time later, the container of water and metal stabilizes at a new temperature of 78°C. Identify the metal.

40. An experiment finds that the specific heat at constant volume of a monatomic gas is 625 J/kg K. Identify the gas.

41. A 150 L (≈ 40 gal) electric hot-water tank has a 5.0 kW heater. How many minutes will it take to raise the water temperature from 65°F to 140°F?

42. An experiment measures the temperature of a 500 g substance while steadily supplying heat to it. Figure P17.42 shows the results of the experiment. What are the (a) specific heat of the solid phase, (b) specific heat of the liquid phase, (c) melting and boiling temperatures, and (d) heats of fusion and vaporization?

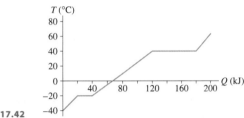

FIGURE P17.42

43. An experiment measures the temperature of a 200 g substance while steadily supplying heat to it. Figure P17.43 shows the results of the experiment. What are (a) the specific heat of the liquid phase and (b) the heat of vaporization?

FIGURE P17.43

44. Liquid nitrogen is used in many low-temperature experiments. It is widely available, and cheaper than gasoline! How much heat must be removed from room temperature (20°C) nitrogen gas to produce 1.0 L of liquid nitrogen? The density of liquid nitrogen is 810 kg/m³.

45. Your 300 mL cup of coffee is too hot to drink when served at 90°C. What is the mass of an ice cube, taken from a −20°C freezer, that will cool your coffee to a pleasant 60°C?

46. A typical nuclear reactor generates 1000 MW (1000 MJ/s) of electrical energy. In doing so, it produces 2000 MW of "waste heat" that must be removed from the reactor to keep it from melting down. Many reactors are sited next to large bodies of water so that they can use the water for cooling. Consider a reactor where the intake water is at 18°C. State regulations limit the temperature of the output water to 30°C so as not to harm aquatic organisms. How many liters of cooling water have to be pumped through the reactor each minute?

47. You find an empty cooler, the kind used to keep drinks cold, at a Fourth of July picnic. The cooler has aluminum walls surrounded by insulating material. This is a 20 L cooler that uses 2.0 kg of aluminum. Just for fun, you toss in a firecracker, slam the lid, and sit on it to keep the lid from blowing off. A minute later, when you open the lid, you see that a built-in thermometer has risen from 25°C to 28°C. How much energy was released by the firecracker when it exploded?

48. A beaker with a metal bottom is filled with 20 g of water at 20°C. It is brought into good thermal contact with a 4000 cm³ container holding 0.40 mol of a monatomic gas at 10 atm pressure. Both containers are well insulated from their surroundings. What is the gas pressure after a long time has elapsed? You can assume that the containers themselves are nearly massless and do not affect the outcome.

FIGURE P17.48

49. The average speed of molecules in 1.0 g of hydrogen gas is 700 m/s.
    a. What is the total microscopic translational kinetic energy of gas?
    b. 500 J of heat are added to the gas, causing it to expand and do 300 J of work. Afterwards, what is the average molecular speed?

50. 2.0 mol of gas are at 30°C and a pressure of 1.5 atm. How much work must be done on the gas to compress it to one third of its initial volume at (a) constant temperature and (b) constant pressure? (c) Show both processes on a single pV diagram.

51. 500 J of work must be done to compress a gas to half its initial volume at constant temperature. How much work must be done to compress the gas by a factor of 10, starting from its initial volume?

52. A cylinder with a 16-cm-diameter piston contains gas at a pressure of 3.0 atm.
    a. How much force does the gas exert on the piston?
    b. How much force does the environment exert on the piston?
    c. The gas expands at constant pressure and pushes the piston out 10 cm. How much work is done by the environment?
    d. How much work is done by the gas?
    e. The thermal energy of the gas increases by 196 J in the expansion of part c. Was heat energy transferred to or from the gas in this process? How much?

53. A 10-cm-diameter cylinder contains argon gas at 10 atm pressure and a temperature of 50°C. A piston can slide in and out of the cylinder. The cylinder's initial length is 20 cm. 2500 J of heat are transferred to the gas, causing the gas to expand at constant pressure. What are (a) the final temperature and (b) the final length of the cylinder?

54. A cube 20 cm on each side contains 3.0 g of helium at 20°C. 1000 J of heat energy are transferred to this gas. What are (a) the final pressure if the process is at constant volume and (b) the final volume if the process is at constant pressure? (c) Show and label both processes on a single pV diagram.

55. 2.0 mol of an ideal monatomic gas are at a pressure of 1.0 atm and a temperature of 310 K. What is the gas's thermal energy? **Hint:** What is the thermal energy at $T = 0$ K?

56. n moles of an ideal gas at temperature T_1 and volume V_1 expand isothermally until the volume has doubled. In terms of n, T_1, and V_1, what are (a) the final temperature, (b) the work done on the gas, and (c) the heat energy transferred to the gas?

57. 5.0 g of nitrogen gas at 20°C and an initial pressure of 3.0 atm undergo an isobaric expansion until the volume has tripled.
    a. What is the gas volume and temperature after the expansion?
    b. How much heat energy is transferred to the gas to cause this expansion?
    The gas pressure is then decreased at constant volume until the original temperature is reached.
    c. What is the gas pressure after the decrease?
    d. What amount of heat energy is transferred from the gas as its pressure decreases?
    e. Show the total process on a pV diagram. Provide an appropriate scale on both axes.

58. Figure P17.58 shows two processes that take a gas from state i to state f. Show that $Q_A - Q_B = p_iV_i$.

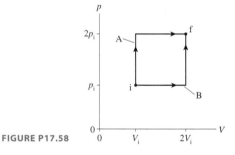

FIGURE P17.58

59. 0.10 mol of nitrogen gas follow the two processes shown in Figure P17.59. How much heat is required for each?

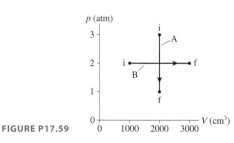

FIGURE P17.59

60. 0.10 mol of nitrogen gas follow the two processes shown in Figure P17.60. How much heat is required for each?

FIGURE P17.60

61. 0.10 mol of a monatomic gas follows the process shown in Figure P17.61.
    a. How much heat energy is transferred to or from the gas during process $1 \rightarrow 2$?
    b. How much heat energy is transferred to or from the gas during process $2 \rightarrow 3$?
    c. What is the total change in thermal energy of the gas?

FIGURE P17.61

62. Two 800 cm³ containers hold identical amounts of a monatomic gas at 20°C. Container A is rigid. Container B has a 100 cm² piston with a mass of 10 kg that can slide up and down vertically without friction. Both containers are placed on identical heaters and heated for equal amounts of time.
    a. Will the final temperature of the gas in A be greater than, less than, or equal to the final temperature of the gas in B? Explain.
    b. Show both processes on a single pV diagram.
    c. What are the initial pressures in containers A and B?
    d. Suppose the heaters have 25 W of power and are turned on for 15 s. What is the final volume of container B?

63. Two cylinders each contain 0.10 mol of a diatomic gas at 300 K and a pressure of 3.0 atm. Cylinder A expands isothermally and cylinder B expands adiabatically until the pressure of each is 1.0 atm.
    a. What are the final temperature and volume of each?
    b. Show both processes on a single pV diagram. Use an appropriate scale on both axes.

64. A monatomic gas follows the process $1 \rightarrow 2 \rightarrow 3$ shown in Figure P17.64. How much heat is needed for (a) process $1 \rightarrow 2$ and (b) process $2 \rightarrow 3$?

FIGURE P17.64

65. Figure P17.65 shows a thermodynamic process followed by 0.015 mol of hydrogen.
    a. How much work is done on the gas?
    b. By how much does the thermal energy of the gas change?
    c. How much heat energy is transferred to the gas?

FIGURE P17.65

66. Figure P17.66 shows a thermodynamic process followed by 120 mg of helium.
    a. Determine the pressure (in atm), temperature (in °C), and volume (in cm³) of the gas at points 1, 2, and 3. Put your results in a table for easy reading.
    b. How much work is done on the gas during each of the three segments?
    c. How much heat energy is transferred to or from the gas during each of the three segments?

FIGURE P17.66

67. One cylinder in the diesel engine of a truck has an initial volume of 600 cm³. Air is admitted to the cylinder at 30°C and a pressure of 1.0 atm. The piston rod then does 400 J of work to rapidly compress the air. What are its final temperature and volume?

68. The volume of a gas is halved during an adiabatic compression that increases the pressure by a factor of 2.5.
    a. What is the specific heat ratio γ?
    b. By what factor does the temperature increase?

69. a. What compression ratio V_{max}/V_{min} will raise the air temperature from 20°C to 1000°C in an adiabatic process?
    b. What pressure ratio p_{max}/p_{min} does this process have?

70. 2.0 g of helium at an initial temperature of 100°C and an initial pressure of 1.0 atm undergo an isobaric expansion until the volume has doubled. What are (a) the final temperature, (b) the work done on the gas, (c) the heat input to the gas, and (d) the change in thermal energy of the gas? (e) Show the process on a pV diagram, using proper scales on both axes.

71. 2.0 g of helium at an initial temperature of 100°C and an initial pressure of 1.0 atm undergo an isothermal expansion until the volume has doubled. What are (a) the final pressure, (b) the work done on the gas, (c) the heat input to the gas, and (d) the change in thermal energy of the gas? (e) Show the process on a pV diagram, using proper scales on both axes.

72. 14.0 g of nitrogen gas at STP are adiabatically compressed to a pressure of 20 atm. What are (a) the final temperature, (b) the work done on the gas, (c) the heat input to the gas, and (d) the compression ratio V_{max}/V_{min}? (e) Show the process on a pV diagram, using proper scales on both axes.

73. 14.0 g of nitrogen gas at STP are compressed in an isochoric process to a pressure of 20 atm. What are (a) the final temperature, (b) the work done on the gas, (c) the heat input to the gas, and (d) the pressure ratio p_{max}/p_{min}? (e) Show the process on a pV diagram, using proper scales on both axes.

74. When strong winds rapidly carry air down from mountains to a lower elevation, the air has no time to exchange heat with its surroundings. The air is compressed as the pressure rises, and its temperature can increase dramatically. These warm winds are called Chinook winds in the Rocky Mountains and Santa Ana winds in California. Suppose the air temperature high in the mountains behind Los Angeles is 0°C at an elevation where the air pressure is 60 kPa. What will the air temperature be, in °C and °F, when the Santa Ana winds have carried this air down to an elevation near sea level where the air pressure is 100 kPa?

In Problems 75 through 78 you are given the equation used to solve a problem. For each of these, you are to
    a. Write a realistic problem for which this is the correct equation.
    b. Finish the solution of the problem.

75. $50 \text{ J} = -n(8.31 \text{ J/mol K})(350 \text{ K})\ln\left(\frac{1}{3}\right)$

76. $(200 \times 10^{-6} \text{ m}^3)(13,600 \text{ kg/m}^3)$
    $\times (140 \text{ J/kg K})(90°\text{C} - 15°\text{C})$
    $+ (0.50 \text{ kg})(449 \text{ J/kg K})(90°\text{C} - T_i) = 0$

77. $-200,000 \text{ J} = \dfrac{M \times 1000 \text{ g/kg}}{28 \text{ g/mol}}(29.1 \text{ J/mol K})$
    $\times (-196°\text{C} - 20°\text{C}) - M(199,000 \text{ J/kg})$

78. $(10 \text{ atm})V_2^{1.40} = (1.0 \text{ atm})V_1^{1.40}$

Challenge Problems

79. Figure CP17.79 shows a thermodynamic process followed by 120 mg of helium.
    a. Determine the pressure (in atm), temperature (in °C), and volume (in cm³) of the gas at points 1, 2, and 3. Put your results in a table for easy reading.
    b. How much work is done on the gas during each of the three segments?
    c. How much heat is transferred to or from the gas during each of the three segments?

FIGURE CP17.79

80. A 6.0-cm-diameter cylinder of nitrogen gas has a 4.0-cm-thick movable copper piston. The cylinder is oriented vertically, as shown in Figure CP17.80, and the air above the piston is evacuated. When the gas temperature is 20°C, the piston floats 20 cm above the bottom of the cylinder.
    a. What is the gas pressure?
    b. How many gas molecules are in the cylinder?
    c. If the average molecular speed is 550 m/s, what is the thermal energy of the gas?
    Then 2.0 J of heat energy are transferred to the gas.
    d. What is the new equilibrium temperature of the gas?
    e. What is the final height of the piston?
    f. How much work is done on the gas as the piston rises?

FIGURE CP17.80

81. You come into lab one day and find a well-insulated 2000 mL thermos bottle containing 500 mL of boiling liquid nitrogen. The remainder of the thermos has nitrogen gas at a pressure of 1.0 atm. The gas and liquid are in thermal equilibrium. While waiting for lab to start, you notice a piece of iron on the table with "197 g" written on it. Just for fun, you drop the iron into the thermos and seal the cap tightly so that no gas can escape. After a few seconds have passed, what is the pressure inside the thermos? The density of liquid nitrogen is 810 kg/m³.

STOP TO THINK ANSWERS

Stop to Think 17.1: a. The piston does work W on the gas. There's no heat, because of the insulation, and $\Delta E_{mech} = 0$ because the gas as a whole doesn't move. Thus $\Delta E_{th} = W > 0$. The work increases the system's thermal energy and thus raises its temperature.

Stop to Think 17.2: d. $W_A = 0$ because A is an isochoric process. $W_B = W_{1\,to\,2} + W_{2\,to\,3}$. $|W_{2\,to\,3}| > |W_{1\,to\,2}|$ because there's more area under the curve, and $W_{2\,to\,3}$ is positive whereas $W_{1\,to\,2}$ is negative. Thus W_B is positive.

Stop to Think 17.3: b and e. The temperature rises in (d) from doing work on the gas ($\Delta E_{th} = W$), not from heat. (e) involves heat because there is a temperature difference. The temperature of the gas doesn't change because the heat is used to do the work of lifting a weight.

Stop to Think 17.4: d. The temperature increases so E_{th} must increase. W is negative in an expansion, so Q must be positive and larger than $|W|$.

Stop to Think 17.5: a. A has a smaller specific heat and thus less thermal inertia. The temperature of A will change more than the temperature of B.

Stop to Think 17.6: a. $W_A + Q_A = W_B + Q_B$. The area under process A is larger than the area under B, so W_A is *more negative* than W_B. Q_A has to be more positive than Q_B to maintain the equality.

18 The Micro/Macro Connection

Heating the air in a hot-air balloon increases the thermal energy of the air molecules. This causes the gas to expand, lowering its density and allowing the balloon to float in the cooler surrounding air.

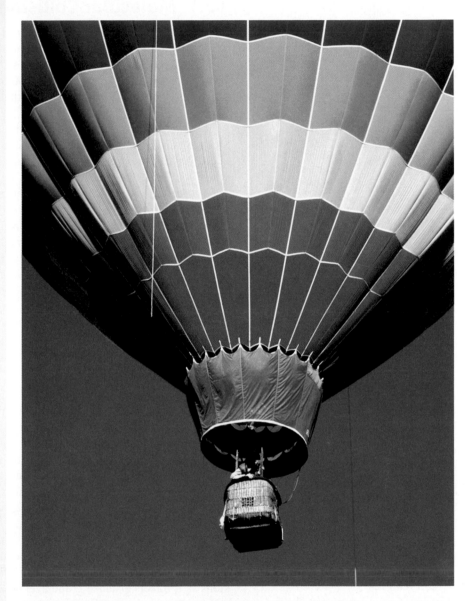

▶ Looking Ahead

The goal of Chapter 18 is to understand the properties of a macroscopic system in terms of the microscopic behavior of its molecules. In this chapter you will learn to:

- Understand how molecular motions and collisions are responsible for macroscopic phenomena such as pressure and heat transfer.
- Establish a connection between temperature, thermal energy, and the average translational kinetic energy of the molecules in the system.
- Use the micro/macro connection to predict the molar specific heats of gases and solids.
- Use the second law of thermodynamics to understand how interacting systems come to thermal equilibrium.

◀ Looking Back

The material in this chapter depends on understanding heat, thermal energy, and the properties of ideal gases. Please review:

- Sections 16.5–16.6 Ideal gases.
- Sections 17.3–17.4 Heat and the first law of thermodynamics.
- Sections 17.5 and 17.7 Specific heats and molar specific heats.

A gas consists of a vast number of molecules ceaselessly colliding with each other and the walls of their container as they whiz about. A solid contains uncountable atoms vibrating around their equilibrium positions. Our goal in this chapter is to show how this turmoil at the microscopic level gives rise to predictable and steady values of macroscopic variables such as pressure, temperature, and specific heat.

This micro/macro connection, which goes by the more formal name **kinetic theory,** will help us elucidate several puzzles that we noted in the previous two chapters. For example, why do all elemental solids have the same molar specific heats, as do all monatomic gases and all diatomic gases? Kinetic theory will also give us a better understanding of *heat* and of how it is that two systems come to thermal equilibrium as they interact.

We'll also introduce a new law of nature, the second law of thermodynamics. The second law is rather subtle, but it has profound implications. We'll use the second law to understand why it is that heat energy "flows" from hot to cold rather than from cold to hot.

18.1 Molecular Collisions

Let's begin by thinking about gases at the atomic level. Chapter 16 presented experimental evidence that the molecules of a gas are moving with different speeds. Those data, seen again in Figure 18.1, show that the molecular speeds in N_2 at room temperature vary from ≈ 100 m/s to ≈ 1200 m/s with the most likely speed being ≈ 550 m/s. You can see that 16% of the molecules have speeds between 600 m/s and 700 m/s.

If you were to repeat the experiment a few seconds or a few hours later, you would again find that the most likely speed is ≈ 550 m/s and that 16% of the molecules have speeds between 600 m/s and 700 m/s. This is somewhat surprising when you think that each molecule in the gas undergoes millions of collisions every second (we'll derive a value later in the chapter). As Figure 18.2 shows, each collision causes one molecule to speed up and the other to slow down.

The "molecular deck of cards" is constantly being reshuffled, yet 16% of the molecules always have speeds between 600 m/s and 700 m/s. There's an important lesson here. We can't possibly keep track of what the individual molecules are doing, but *averages,* such as the average number of molecules in the speed range 600 to 700 m/s, can have precise, predictable values. **The micro/macro connection is built on the idea that macroscopic properties of a system, such as temperature or pressure, are related to the *average* behavior of the atoms and molecules.**

Mean Free Path

Imagine someone opening a bottle of strong perfume a few feet away from you. If molecular speeds are hundreds of meters per second, you might expect to smell the perfume almost instantly. But that isn't what happens. As you know, it takes many seconds for the molecules to *diffuse* across the room. Let's see why this is.

Figure 18.3 shows a "movie" of one molecule as it moves through a gas. Instead of zipping along in a straight line, as it would in a vacuum, the molecule follows a convoluted zig-zag path in which it frequently collides with other molecules. A molecule may have traveled hundreds of meters by the time it manages to get 1 or 2 m away from its starting point.

The random distribution of the molecules in the gas causes the straight-line segments between collisions to be of unequal lengths. A question we could ask is, "What is the *average* distance between collisions?" If a molecule has N_{coll} collisions as it travels distance L, the average distance between collisions, which is called the **mean free path** λ (lowercase Greek lambda), is

$$\lambda = \frac{L}{N_{coll}} \tag{18.1}$$

The concept of mean free path is used not only in gases but also to describe electrons moving through conductors and light passing through a medium that scatters the photons.

Our task is to determine the number of collisions. Figure 18.4a shows two molecules approaching each other. We will assume that the molecules are spherical and of radius r. We will also continue the ideal-gas assumption that the molecules undergo hard-sphere collisions, like billiard balls. In that case, the molecules will

FIGURE 18.1 The distribution of molecular speeds in a sample of nitrogen gas.

FIGURE 18.2 Frequent collisions change the speed and direction of molecules in a gas.

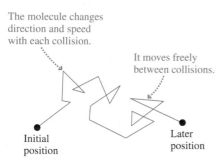

FIGURE 18.3 A single molecule follows a zig-zag path through a gas as it collides with other molecules.

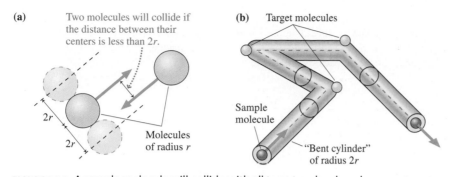

FIGURE 18.4 A sample molecule will collide with all target molecules whose centers are within a bent cylinder of radius $2r$ centered on its path.

collide if the distance between their *centers* is less than $2r$. They will miss if the distance is greater than $2r$.

Figure 18.4b shows a cylinder of radius $2r$ centered on the trajectory of a "sample" molecule. The sample molecule collides with any "target" molecule whose center is located within the cylinder, causing the cylinder to bend at that point. Hence the number of collisions N_{coll} is equal to the number of molecules in a cylindrical volume of length L.

The volume of a cylinder is $V_{cyl} = AL = \pi(2r)^2 L$. If the number density of the gas is N/V particles per m³, then the number of collisions along a trajectory of length L is

$$N_{coll} = \frac{N}{V}V_{cyl} = \frac{N}{V}\pi(2r)^2 L = 4\pi\frac{N}{V}r^2 L \qquad (18.2)$$

Thus the mean free path between collisions is

$$\lambda = \frac{L}{N_{coll}} = \frac{1}{4\pi(N/V)r^2}$$

We made a tacit assumption in this derivation that the target molecules are at rest. While the general idea behind our analysis is correct, a more careful calculation in which all molecules move introduces an extra factor of $\sqrt{2}$, giving

$$\lambda = \frac{1}{4\sqrt{2}\pi(N/V)r^2} \qquad \text{(mean free path)} \qquad (18.3)$$

Measurements are necessary to determine precise values of atomic and molecular radii, but a reasonable rule of thumb is to assume that atoms in a monatomic gas have $r \approx 0.5 \times 10^{-10}$ m and diatomic molecules have $r \sim 1.0 \times 10^{-10}$ m.

EXAMPLE 18.1 The mean free path at STP
What is the mean free path of a nitrogen molecule at 1.0 atm pressure and room temperature (20°C)?

SOLVE Nitrogen is a diatomic molecule, so $r \approx 1.0 \times 10^{-10}$ m. We can use the ideal-gas law in the form $pV = Nk_BT$ to determine the number density:

$$\frac{N}{V} = \frac{p}{k_BT} = \frac{101{,}300 \text{ Pa}}{(1.38 \times 10^{-23} \text{ J/K})(293 \text{ K})} = 2.51 \times 10^{25} \text{ m}^{-3}$$

Thus the mean free path is

$$\lambda = \frac{1}{4\sqrt{2}\pi(N/V)r^2}$$

$$= \frac{1}{4\sqrt{2}\pi(2.51 \times 10^{25} \text{ m}^{-3})(1.0 \times 10^{-10} \text{ m})^2}$$

$$= 2.25 \times 10^{-7} \text{ m} = 225 \text{ nm}$$

ASSESS You learned in Example 16.5 that the average separation between gas molecules at STP is ≈5.7 nm. It seems that any given molecule can slip between its neighbors, which are spread out in three dimensions, and travel—on average—about 40 times the average spacing before it collides with another molecule.

STOP TO THINK 18.1 The table shows the properties of four gases, each having the same number of molecules. Rank in order, from largest to smallest, the mean free paths λ_A to λ_D of molecules in these gases.

Gas	A	B	C	D
Volume	V	$2V$	V	V
Atomic mass	m	m	$2m$	m
Atomic radius	r	r	r	$2r$

18.2 Pressure in a Gas

Why does a gas have pressure? In Chapter 15, where pressure was introduced, we suggested that the pressure in a gas is due to collisions of the molecules with walls of its container. The force due to one such collision may be unmeasurably tiny, but the steady rain of a vast number of molecules striking a wall each second exerts a measurable macroscopic force. The gas pressure is the force per unit area ($p = F/A$) resulting from these molecular collisions.

Our task in this section is to calculate the pressure by doing the appropriate averaging over molecular motions and collisions. This task can be divided into three main pieces:

1. Calculate the force a single molecule exerts on the wall during a collision.
2. Add the force due to all collisions.
3. Introduce an appropriate average speed.

Force Due to a Single Collision

Figure 18.5 shows a molecule with an x-component of velocity v_x approaching a wall. We will assume that the collision with the wall is perfectly elastic, an assumption we will justify later, in which case the molecule rebounds from the wall with its x-component of velocity changed from $+v_x$ to $-v_x$. This molecule experiences an impulse. We can use the impulse-momentum theorem from Chapter 9 to write

$$(J_x)_{\text{molecule}} = \Delta p = m(-v_x) - mv_x = -2mv_x \tag{18.4}$$

According to Newton's third law, the wall experiences the equal but opposite impulse

$$(J_x)_{\text{wall}} = +2mv_x \tag{18.5}$$

as a result of this single collision.

Impulse was defined in Chapter 9 as

$$J_x = \int_0^{\Delta t_{\text{coll}}} F_x(t)\, dt \tag{18.6}$$

where $F_x(t)$ is the time-varying x-component of the force the molecule exerts on the wall during the collision and Δt_{coll} is the duration of the collision. A typical collision force is shown in Figure 18.6a.

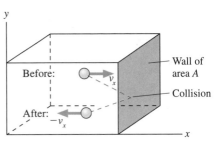

FIGURE 18.5 A molecule colliding with the wall exerts an impulse J_x on it.

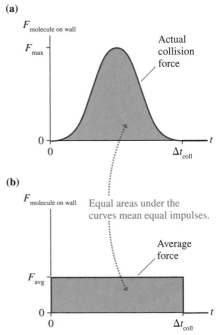

FIGURE 18.6 The force that a molecule exerts on the wall during a collision.

Suppose we replace the actual collision force with an *average* force F_{avg} defined, as Figure 18.6b shows, such that the area under the curve, or the impulse, is the same as for the true force. Then

$$(J_x)_{\text{wall}} = \int_0^{\Delta t_{\text{coll}}} F_{\text{avg}} \, dt = F_{\text{avg}} \Delta t_{\text{coll}} = 2mv_x$$

Thus the average force on the wall during *one* collision is

$$F_{\text{avg}} = \frac{2mv_x}{\Delta t_{\text{coll}}} \tag{18.7}$$

EXAMPLE 18.2 The average force in a collision
Estimate the average force a nitrogen molecule exerts on the wall of its container during a collision.

SOLVE We saw in the previous section that a typical molecular speed is ≈500 m/s. Although we don't know the details, the duration of a collision must be roughly the time it takes a molecule to move a distance equal to its radius. We noted that the radius of a typical diatomic molecule is 1.0×10^{-10} m, so a rough estimate for the duration of a collision is

$$\Delta t_{\text{coll}} \approx \frac{1.0 \times 10^{-10} \text{ m}}{500 \text{ m/s}} \approx 2 \times 10^{-13} \text{ s}$$

A nitrogen molecule has mass $m = 28$ u $= 4.65 \times 10^{-26}$ kg. Thus an estimate of the average force during a collision is

$$F_{\text{avg}} = \frac{2mv_x}{\Delta t_{\text{coll}}} \approx \frac{2(4.65 \times 10^{-26} \text{ kg})(500 \text{ m/s})}{2 \times 10^{-13} \text{ s}} \approx 2 \times 10^{-10} \text{ N}$$

ASSESS The duration of a collision is extremely short, and the force exerted by one molecule is very tiny. But a very large number of molecules are all colliding essentially simultaneously.

Adding the Forces

The net force exerted on the wall by *all* the molecules is the average force per molecule, given by Equation 18.7, multiplied by the total number N_{coll} of such collisions that occur during the time interval Δt_{coll}. That is,

$$F_{\text{net}} = N_{\text{coll}} F_{\text{avg}} = 2mv_x \frac{N_{\text{coll}}}{\Delta t_{\text{coll}}} \tag{18.8}$$

The ratio $N_{\text{coll}}/\Delta t_{\text{coll}}$ is the *rate* at which molecules collide with the wall.

To find N_{coll}, the number of molecules that hit the wall during the time interval Δt_{coll}, assume for the moment that all molecules have the *same* x-component velocity v_x. These molecules all travel distance $\Delta x = v_x \Delta t_{\text{coll}}$ along the x-axis during interval Δt_{coll}. Figure 18.7 has shaded the gas that is within distance Δx of the wall. *Every one* of the molecules in this shaded region that is moving to the right will reach and collide with the wall during time Δt_{coll}. Molecules outside this region will not reach the wall during Δt_{coll} because they are not moving fast enough to travel distance Δx.

The shaded region has volume $A\Delta x$, where A is the surface area of the wall. Only half the molecules are moving to the right, hence the number of collisions during Δt_{coll} is

$$N_{\text{coll}} = \frac{1}{2}\frac{N}{V} A\Delta x = \frac{1}{2}\frac{N}{V} Av_x \Delta t_{\text{coll}} \tag{18.9}$$

Thus the *rate* of molecular collisions with a wall of area A is

$$\text{rate of collisions} = \frac{N_{\text{coll}}}{\Delta t_{\text{coll}}} = \frac{1}{2}\frac{N}{V} Av_x \tag{18.10}$$

Only molecules moving to the right in the shaded region will hit the wall during Δt_{coll}.

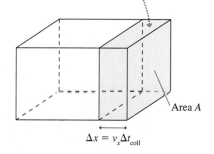

Area A

$\Delta x = v_x \Delta t_{\text{coll}}$

FIGURE 18.7 Determining the number of collisions.

The net force on the wall is found by substituting $N_{coll}/\Delta t_{coll}$ from Equation 18.10 into Equation 18.8:

$$F_{net} = (2mv_x)\left(\frac{1}{2}\frac{N}{V}Av_x\right) = \frac{N}{V}mv_x^2A \qquad (18.11)$$

Notice that this expression for F_{net} does not depend on any details of the molecular collision.

Equation 18.11 is based on an assumption that all molecules have the same speed. This is a condition we need to relax if our result is to apply to real gases. We can correct for this by replacing the squared velocity v_x^2 in Equation 18.11 with its average value. That is,

$$F_{net} = \frac{N}{V}m(v_x^2)_{avg}A \qquad (18.12)$$

where $(v_x^2)_{avg}$ is the quantity v_x^2 averaged over all molecules in the container.

The Root-Mean-Square Speed

We need to be somewhat careful when averaging velocities. The velocity component v_x has a sign. At any instant of time, half the molecules in a container move to the right and have positive v_x while the other half move to the left and have negative v_x. Thus the *average velocity* is $(v_x)_{avg} = 0$. If this weren't true, the entire container of gas would move away!

The speed of a molecule is $v = (v_x^2 + v_y^2 + v_z^2)^{1/2}$. Thus the average of the speed squared is

$$(v^2)_{avg} = (v_x^2 + v_y^2 + v_z^2)_{avg} = (v_x^2)_{avg} + (v_y^2)_{avg} + (v_z^2)_{avg} \qquad (18.13)$$

The square root of $(v^2)_{avg}$ is called the **root-mean-square speed** v_{rms}:

$$v_{rms} = \sqrt{(v^2)_{avg}} \qquad \text{(root-mean-square speed)} \qquad (18.14)$$

This is usually called the *rms speed*. You can remember its definition by noting that its name is the *opposite* of the sequence of operations: first you square all the speeds, then you average the squares (find the mean), then you take the square root. Because the square root "undoes" the square, v_{rms} must, in some sense, give an average speed.

NOTE ▶ We could compute a true average speed v_{avg}, but that calculation turns out to be difficult. More important, the root-mean-square speed tends to arise naturally in many scientific and engineering calculations. It turns out that v_{rms} differs from v_{avg} by less than 10%, so for practical purposes we can interpret v_{rms} as being essentially the average speed of a molecule in a gas. ◀

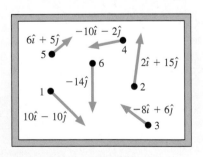

FIGURE 18.8 The molecular velocities of Example 18.3. Units are m/s.

EXAMPLE 18.3 Calculating the root-mean-square speed

Figure 18.8 shows the velocities of all the molecules in a six-molecule two-dimensional gas. Calculate and compare the average velocity $\vec{v}_{avg}$, the average speed v_{avg}, and the rms speed v_{rms}.

SOLVE Table 18.1 shows the velocity components v_x and v_y for each molecule, the squares v_x^2 and v_y^2, their sum $v^2 = v_x^2 + v_y^2$, and the speed $v = (v_x^2 + v_y^2)^{1/2}$. Averages of all the values in each column are shown at the bottom. You can see that the average velocity is $\vec{v}_{avg} = \vec{0}$ m/s and the average speed is $v_{avg} = 11.9$ m/s. The rms speed is

$$v_{rms} = \sqrt{(v^2)_{avg}} = \sqrt{148.3 \text{ m}^2/\text{s}^2} = 12.2 \text{ m/s}$$

ASSESS The rms speed is only 2.5% greater than the average speed.

TABLE 18.1 Calculation of rms speed and average speed for the molecules of Example 18.3

Molecule	v_x	v_y	v_x^2	v_y^2	v^2	v
1	10	−10	100	100	200	14.1
2	2	15	4	225	229	15.1
3	−8	6	64	36	100	10.0
4	−10	−2	100	4	104	10.2
5	6	5	36	25	61	7.8
6	0	−14	0	196	196	14.0
Average	0	0			148.3	11.9

There's nothing special about the *x*-axis. The coordinate system is something that *we* impose on the problem, so *on average* it must be the case that

$$(v_x^2)_{avg} = (v_y^2)_{avg} = (v_z^2)_{avg} \qquad (18.15)$$

Hence we can use Equation 18.13 and the definition of v_{rms} to write

$$v_{rms}^2 = (v_x^2)_{avg} + (v_y^2)_{avg} + (v_z^2)_{avg} = 3(v_x^2)_{avg}$$

Consequently, $(v_x^2)_{avg}$ is

$$(v_x^2)_{avg} = \frac{1}{3}v_{rms}^2 \qquad (18.16)$$

Using this result in Equation 18.12 gives us the net force on the wall of the container:

$$F_{net} = \frac{1}{3}\frac{N}{V}mv_{rms}^2 A \qquad (18.17)$$

Thus the pressure on the wall of the container due to all the molecular collisions is

$$p = \frac{F}{A} = \frac{1}{3}\frac{N}{V}mv_{rms}^2 \qquad (18.18)$$

We have met our goal. Equation 18.18 expresses the macroscopic pressure in terms of the microscopic physics. The pressure depends on the number density of molecules in the container and on how fast, on average, the molecules are moving.

EXAMPLE 18.4 The rms speed of helium atoms
A container holds helium at a pressure of 200 kPa and a temperature of 60°C. What is the rms speed of the helium atoms?

SOLVE The rms speed can be found from the pressure and the number density. Using the ideal-gas law gives us the number density:

$$\frac{N}{V} = \frac{p}{k_B T} = \frac{200,000 \text{ Pa}}{(1.38 \times 10^{-23} \text{ J/K})(333 \text{ K})} = 4.35 \times 10^{25} \text{ m}^{-3}$$

The mass of a helium atom is $m = 4\text{ u} = 6.64 \times 10^{-27}$ kg. Thus

$$v_{rms} = \sqrt{\frac{3p}{(N/V)m}} = 1440 \text{ m/s}$$

The speed of every molecule in a gas is suddenly increased by a factor of 4. As a result, v_{rms} increases by a factor of

a. 2.
c. 4.
e. 16.

b. <4, but not necessarily 2.
d. >4 but not necessarily 16.
f. v_{rms} doesn't change.

18.3 Temperature

A molecule of mass m and velocity v has translational kinetic energy

$$\epsilon = \frac{1}{2}mv^2 \tag{18.19}$$

We'll use ϵ (lowercase Greek epsilon) to distinguish the energy of a molecule from the system energy E. Thus the average translational kinetic energy is

$$
\begin{aligned}
\epsilon_{avg} &= \text{average translational kinetic energy per molecule} \\
&= \frac{1}{2}m(v^2)_{avg} = \frac{1}{2}mv_{rms}^2
\end{aligned}
\tag{18.20}
$$

We've included the word *translational* to distinguish ϵ from rotational kinetic energy, which we will consider later in this chapter.

We can write the gas pressure, Equation 18.18, in terms of the average translational kinetic energy as

$$p = \frac{2}{3}\frac{N}{V}\left(\frac{1}{2}mv_{rms}^2\right) = \frac{2}{3}\frac{N}{V}\epsilon_{avg} \tag{18.21}$$

The pressure is directly proportional to the average molecular translational kinetic energy. This makes sense. More-energetic molecules will hit the walls harder as they bounce and thus exert more force on the walls.

It's instructive to write Equation 18.21 as

$$pV = \frac{2}{3}N\epsilon_{avg} \tag{18.22}$$

We know, from the ideal-gas law, that

$$pV = Nk_{B}T \tag{18.23}$$

Comparing these two equations, we reach the significant conclusion that the average translational kinetic energy per molecule is

$$\epsilon_{avg} = \frac{3}{2}k_{B}T \quad \text{(average translational kinetic energy)} \tag{18.24}$$

where the temperature T is in kelvins. For example, the average translational kinetic energy of a molecule at room temperature (20°C) is

$$\epsilon_{avg} = \frac{3}{2}(1.38 \times 10^{-23} \text{ J/K})(293 \text{ K}) = 6.1 \times 10^{-21} \text{ J}$$

NOTE ▶ The average translational kinetic energy per molecule depends *only* on the temperature, not on the molecule's mass. If two gases have the same temperature, their molecules have the same average translational kinetic energy. This will be an important idea when we look at the thermal interaction between two systems. ◀

Equation 18.24 is especially satisfying because it finally gives real meaning to the concept of temperature. Writing it as

$$T = \frac{2}{3k_B}\epsilon_{avg} \qquad (18.25)$$

we can see that, for a gas, **this thing we call *temperature* is a direct measure of the average translational kinetic energy.** A higher temperature corresponds to a larger value of ϵ_{avg} and thus to higher molecular speeds. This concept of temperature also gives meaning to *absolute zero* as the temperature at which $\epsilon_{avg} = 0$ and all molecular motion ceases. (Quantum effects at very low temperatures prevent the motions from actually stopping, but our classical theory predicts that they would.) Figure 18.9 summarizes what we've learned thus far about the micro/macro connection.

We can now justify our assumption that molecular collisions are perfectly elastic. Suppose they were not. That is, suppose that a small amount of kinetic energy was lost in each collision. If that were so, the average translational kinetic energy ϵ_{avg} of the gas would slowly decrease and we would see a steadily decreasing temperature. But that doesn't happen. The temperature of an isolated system remains perfectly constant, indicating that ϵ_{avg} is not changing with time. Consequently, the collisions must be perfectly elastic.

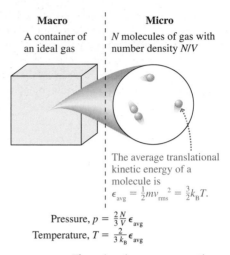

Macro	**Micro**
A container of an ideal gas	N molecules of gas with number density N/V

The average translational kinetic energy of a molecule is
$\epsilon_{avg} = \frac{1}{2}mv_{rms}^2 = \frac{3}{2}k_BT.$

Pressure, $p = \frac{2}{3}\frac{N}{V}\epsilon_{avg}$

Temperature, $T = \frac{2}{3k_B}\epsilon_{avg}$

FIGURE 18.9 The micro/macro connection for pressure and temperature.

EXAMPLE 18.5 Total microscopic kinetic energy
What is the total translational kinetic energy of the molecules in 1.0 mol of gas at STP?

SOLVE The average translational kinetic energy per molecule is

$$\epsilon_{avg} = \frac{3}{2}k_BT = \frac{3}{2}(1.38 \times 10^{-23}\ \text{J/K})(273\ \text{K})$$

$$= 5.65 \times 10^{-21}\ \text{J}$$

1.0 mol of gas contains N_A molecules, hence the total kinetic energy is

$$K_{micro} = N_A\epsilon_{avg} = 3400\ \text{J}$$

ASSESS The energy of any one molecule is incredibly small. Nonetheless, a macroscopic system has substantial thermal energy because it consists of an incredibly large number of molecules.

By definition, $\epsilon_{avg} = \frac{1}{2}mv_{rms}^2$. Using the ideal-gas law, we found $\epsilon_{avg} = \frac{3}{2}k_BT$. By equating these two expressions we find that the rms speed of a molecule is

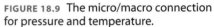

ActivPhysics 8.1–8.3

$$\boxed{v_{rms} = \sqrt{\frac{3k_BT}{m}}} \qquad (18.26)$$

The rms speed of a molecule depends on the square root of the temperature and inversely on the square root of the molecular mass.

EXAMPLE 18.6 Calculating an rms speed
What is the rms speed of a nitrogen molecule at room temperature?

SOLVE The molecular mass is $m = 28\ \text{u} = 4.68 \times 10^{-26}\ \text{kg}$ and $T = 20°\text{C} = 293\ \text{K}$. It is then a simple calculation to find

$$v_{rms} = \sqrt{\frac{3(1.38 \times 10^{-23}\ \text{J/K})(293\ \text{K})}{4.68 \times 10^{-26}\ \text{kg}}} = 509\ \text{m/s}$$

Some speeds will be greater than this and others smaller, but 509 m/s will be a typical or fairly average speed. This is in excellent agreement with the experimental results of Figure 18.1.

EXAMPLE 18.7 Laser cooling

It is possible to "cool" atoms by letting them interact with a laser beam under proper, carefully controlled conditions. Laser cooling is currently a subject of intense research activity, and it is now possible to cool a dilute gas of atoms to a temperature less than one *micro*kelvin! (The atoms are kept from solidifying by their extremely low density.) Various novel quantum effects appear at these incredibly low temperatures. What is the rms speed for a cesium atom at a temperature of 1.0 μK?

SOLVE Reference to the periodic table of the elements shows that the atomic mass of cesium is $m = 133$ u $= 2.22 \times 10^{-25}$ kg. At $T = 1\ \mu$K $= 1 \times 10^{-6}$ K the rms speed is

$$v_{rms} = \sqrt{\frac{3(1.38 \times 10^{-23}\ \text{J/K})(1.0 \times 10^{-6}\ \text{K})}{2.22 \times 10^{-25}\ \text{kg}}}$$

$$= 0.014\ \text{m/s} = 1.4\ \text{cm/s}$$

This is slow enough to enable us to "watch" the atoms moving about!

EXAMPLE 18.8 Mean time between collisions

Because v_{rms} is essentially the average molecular speed, the *mean time between collisions* is simply the time needed to travel distance λ, the mean free path, at speed v_{rms}. What is the mean time between collisions for a nitrogen molecule at 1.0 atm pressure and room temperature?

SOLVE We found $\lambda = 2.25 \times 10^{-7}$ m in Example 18.1 and $v_{rms} = 509$ m/s in Example 18.6. Thus the mean time between collisions is

$$\tau_{coll} = \frac{\lambda}{v_{rms}} = \frac{2.25 \times 10^{-7}\ \text{m}}{509\ \text{m/s}} = 4.42 \times 10^{-10}\ \text{s}$$

ASSESS The air molecules around us move very fast, they collide with their neighbors about two billion times every second, and they manage to move, on average, only about 225 nm between collisions.

STOP TO THINK 18.3   Which system (or systems) has the largest average translational kinetic energy per molecule?

a. 1 mol of He at $p = 1$ atm, $T = 300$ K
b. 2 mol of He at $p = 2$ atm, $T = 300$ K
c. 1 mol of N$_2$ at $p = 0.5$ atm, $T = 600$ K
d. 2 mol of N$_2$ at $p = 0.5$ atm, $T = 450$ K
e. 1 mol of Ar at $p = 0.5$ atm, $T = 450$ K
f. 2 mol of Ar at $p = 2$ atm, $T = 300$ K

18.4 Thermal Energy and Specific Heat

We defined the thermal energy of a system to be $E_{th} = K_{micro} + U_{micro}$, where K_{micro} is the microscopic kinetic energy of the moving molecules and U_{micro} is the potential energy of the stretched and compressed molecular bonds. We're now ready to take a microscopic look at thermal energy. In doing so, we'll be able to resolve the puzzle of the molar specific heats.

Monatomic Gases

Figure 18.10 shows a monatomic gas such as helium or neon. The atoms in an ideal gas have no molecular bonds with their neighbors, hence $U_{micro} = 0$. Furthermore, the kinetic energy of a monatomic gas particle is entirely translational kinetic energy ϵ. Thus the thermal energy of a monatomic gas of N atoms is

$$E_{th} = K_{micro} = \epsilon_1 + \epsilon_2 + \epsilon_3 + \cdots + \epsilon_N = N\epsilon_{avg} \tag{18.27}$$

Atom i has translational kinetic energy ϵ_i but no potential energy or rotational kinetic energy.

N atoms in a gas at temperature T

The thermal energy of the gas is $E_{th} = \epsilon_1 + \epsilon_2 + \epsilon_3 + \cdots = N\epsilon_{avg}$.

FIGURE 18.10 The atoms in a monatomic gas have only translational kinetic energy.

where ϵ_i is the translational kinetic energy of atom i. We found that $\epsilon_{avg} = \frac{3}{2}k_B T$, hence the thermal energy is

$$E_{th} = \frac{3}{2}Nk_B T = \frac{3}{2}nRT \qquad \text{(thermal energy of a monatomic gas)} \qquad (18.28)$$

where we used $N = nN_A$ and the definition of Boltzmann's constant, $k_B = R/N_A$.

We've noted for the last two chapters that thermal energy is associated with temperature. Now we have an explicit result for a monatomic gas: E_{th} is directly proportional to the temperature. Notice that E_{th} is independent of the atomic mass. Any two monatomic gases will have the same thermal energy if they have the same temperature and the same number of atoms (or moles).

If the temperature of a monatomic gas changes by ΔT, its thermal energy changes by

$$\Delta E_{th} = \frac{3}{2}nR\Delta T \qquad (18.29)$$

In Chapter 17 we found that the change in thermal energy for *any* ideal gas process is related to the molar specific heat at constant volume by

$$\Delta E_{th} = nC_V \Delta T \qquad (18.30)$$

Equation 18.29 is a microscopic result that we obtained by relating the temperature to the average translational kinetic energy of the atoms. Equation 18.30 is a macroscopic result that we arrived at from the first law of thermodynamics. We can make a micro/macro connection by combining these two equations. Doing so gives us a *prediction* for the molar specific heat:

$$C_V = \frac{3}{2}R = 12.5 \text{ J/mol K} \qquad \text{(monatomic gas)} \qquad (18.31)$$

This was exactly the value of C_V for all three monatomic gases in Table 17.4. Not only has the micro/macro connection shown that C_V is the same for all monatomic gases—the puzzle we had noted in Chapter 17—it has also made a specific prediction for the *value* of C_V. The perfect agreement of theory and experiment is strong evidence that gases really do consist of moving, colliding molecules.

The Equipartition Theorem

The particles of a monatomic gas are atoms. Their energy consists exclusively of their translational kinetic energy. A particle's translational kinetic energy can be written

$$\epsilon = \frac{1}{2}mv^2 = \frac{1}{2}mv_x^2 + \frac{1}{2}mv_y^2 + \frac{1}{2}mv_z^2 = \epsilon_x + \epsilon_y + \epsilon_z \qquad (18.32)$$

where we have written separately the energy associated with translational motion along the three axes. Because each axis in space is independent, we can think of ϵ_x, ϵ_y, and ϵ_z as independent *modes* of storing energy within the system.

Other systems have additional modes of energy storage. For example,

- Two atoms joined by a spring-like molecular bond can vibrate back and forth. Both kinetic and potential energy are associated with this vibration.
- A diatomic molecule, in additional to translational kinetic energy, has rotational kinetic energy if it rotates end-over-end like a dumbbell.

It will be useful to define the number of **degrees of freedom** as the number of distinct and independent modes of energy storage. A monatomic gas has three degrees of freedom, the three modes of translational kinetic energy. Systems that can vibrate or rotate have more degrees of freedom.

An important result of statistical physics says that the energy in a system is distributed so that all modes of energy storage have equal amounts of energy. This conclusion is known as the *equipartition theorem,* meaning that the energy is equally divided. The proof is beyond what we can do in this textbook, so we will state the theorem without proof:

> **Equipartition theorem** The thermal energy of a system of particles is equally divided among all the possible energy modes. For a system of N particles at temperature T, the energy stored in each mode (each degree of freedom) is $\frac{1}{2}Nk_BT$ or, in terms of moles, $\frac{1}{2}nRT$.

A monatomic gas has three degrees of freedom and thus, as we found above, $E_{th} = \frac{3}{2}Nk_BT$.

Solids

Each atom has microscopic translational kinetic energy *and* microscopic potential energy along all three axes.

FIGURE 18.11 A simple model of a solid.

Figure 18.11 reminds you of our "bedspring model" of a solid with particle-like atoms connected by a lattice of spring-like molecular bonds. How many degrees of freedom does a solid have? The kinetic energy of an atom as it vibrates around its equilibrium position is given by Equation 18.32. Three degrees of freedom are associated with the kinetic energy, just as in a monatomic gas. In addition, the molecular bonds can be compressed or stretched independently along the x-, y-, and z-axes. Three additional degrees of freedom are associated with these three modes of potential energy. Altogether, a solid has six degrees of freedom.

The energy stored in each of these six degrees of freedom is $\frac{1}{2}Nk_BT$. The thermal energy of a solid is the total energy stored in all six modes, or

$$E_{th} = 3Nk_BT = 3nRT \qquad \text{(thermal energy of a solid)} \qquad (18.33)$$

We can use this result to predict the molar specific heat of a solid. If the temperature changes by ΔT, then the thermal energy changes by

$$\Delta E_{th} = 3nR\Delta T \qquad (18.34)$$

In Chapter 17 we defined the molar specific heat of a solid such that

$$\Delta E_{th} = nC\Delta T \qquad (18.35)$$

By comparing Equations 18.34 and 18.35 we can predict that the molar specific heat of a solid is

$$C = 3R = 25.0 \text{ J/mol K} \qquad \text{(solid)} \qquad (18.36)$$

Not bad. The five elemental solids in Table 17.2 had molar specific heats clustered right around 25 J/mol K. They ranged from 24.3 J/mol K for aluminum to 26.5 J/mol K for lead. There are two reasons why the agreement between theory and experiment isn't quite as perfect as it was for monatomic gases. First, our simple bedspring model of a solid isn't quite as accurate as our model of a monatomic gas. Second, quantum effects are beginning to make their appearance. More on this shortly. Nonetheless, our ability to predict C to within a few percent from a simple model of a solid is further evidence for the atomic structure of matter.

Diatomic Molecules

Diatomic molecules are a bigger challenge. How many degrees of freedom does a diatomic molecule have? Figure 18.12 shows a diatomic molecule, such as molecular nitrogen N_2, oriented along the z-axis. Three degrees of freedom are associated with the molecule's translational kinetic energy. The molecule can have a dumbbell-like end-over-end rotation about either the x-axis or the y-axis. It can

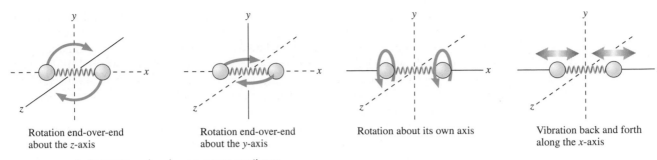

Rotation end-over-end about the *z*-axis

Rotation end-over-end about the *y*-axis

Rotation about its own axis

Vibration back and forth along the *x*-axis

FIGURE 18.12 A diatomic molecule can rotate or vibrate.

also rotate about its own axis. These are three rotational degrees of freedom. The two atoms can also vibrate back and forth, stretching and compressing the molecular bond. This vibrational motion has both kinetic and potential energy, thus two more degrees of freedom.

Altogether, then, a diatomic molecule has eight degrees of freedom, and we would expect the thermal energy of a gas of diatomic molecules to be $E_{th} = 4k_B T$. The analysis we followed for a monatomic gas would then lead to the prediction $C_V = 4R = 33.2$ J/mol K. As compelling as this reasoning seems to be, this is *not* the experimental value of C_V that was reported for diatomic gases in Table 17.4. Instead, we found $C_V = 20.8$ J/mol K.

Why should a theory that works so well for monatomic gases and solids fail so miserably for diatomic molecules? To see what's going on, notice that 20.8 J/mol K = $\frac{5}{2}R$. A monatomic gas, with three degrees of freedom, has $C_V = \frac{3}{2}R$. A solid, with six degrees of freedom, has $C = 3R$. A diatomic gas would have $C_V = \frac{5}{2}R$ if it had five degrees of freedom, not eight.

This discrepancy was a major conundrum as statistical physics developed in the late 19th century. Although it was not recognized as such at the time, we are here seeing our first evidence for the breakdown of classical Newtonian physics. Classically, a diatomic molecule has eight degrees of freedom. The equipartition theorem doesn't distinguish between them; all eight should have the same energy. But atoms and molecules are not classical particles. It took the development of quantum theory in the 1920s to accurately characterize the behavior of atoms and molecules. We don't yet have the tools needed to see why, but quantum effects prevent three of the modes—the two vibrational modes and the rotation of the molecule about its own axis—from being active at room temperature.

Figure 18.13 shows C_V as a function of temperature for hydrogen gas. C_V is right at $\frac{5}{2}R$ for temperatures from ≈200 K up to ≈800 K. But at very low temperatures C_V drops to the monatomic-gas value $\frac{3}{2}R$. The two rotational modes

FIGURE 18.13 Hydrogen molar specific heat at constant volume as a function of temperature. The temperature scale is logarithmic.

become "frozen out" and the nonrotating molecule has only translational kinetic energy. Quantum physics can explain this, but not Newtonian physics. You can also see that the two vibrational modes *do* become active at very high temperatures, where C_V rises to $\frac{7}{2}R$. Thus the real answer to "What's wrong?" is that Newtonian physics is not the right physics for describing atoms and molecules. We are somewhat fortunate that Newtonian physics is adequate to understand monatomic gases and solids, at least at room temperature.

Accepting the quantum result that a diatomic gas has only five degrees of freedom at commonly used temperatures (the translational degrees of freedom and the two end-over-end rotations), we find

$$E_{th} = \frac{5}{2}Nk_BT = \frac{5}{2}nRT$$
(diatomic gases)
$$C_V = \frac{5}{2}R = 20.8 \text{ J/mol K}$$
(18.37)

A diatomic gas has more thermal energy than a monatomic gas at the same temperature because the molecules have rotational as well as translational kinetic energy.

While the micro/macro connection firmly establishes the atomic structure of matter, it also heralds the need for a new theory of matter at the atomic level. That is a task we will take up in Part VII. For now, Table 18.2 summarizes what we have learned from kinetic theory about thermal energy and molar specific heats.

TABLE 18.2 Kinetic theory predictions for the thermal energy and the molar specific heat

System	Degrees of freedom	E_{th}	C_V
Monatomic gas	3	$\frac{3}{2}Nk_BT = \frac{3}{2}nRT$	$\frac{3}{2}R = 12.5$ J/mol K
Diatomic gas	5	$\frac{5}{2}Nk_BT = \frac{5}{2}nRT$	$\frac{5}{2}R = 20.8$ J/mol K
Elemental solid	6	$3Nk_BT = 3nRT$	$3R = 25.0$ J/mol K

EXAMPLE 18.9 The rotational frequency of a molecule
The nitrogen molecule N_2 has a bond length of 0.12 nm. Estimate the rotational frequency of N_2 at 20°C.

MODEL Each mode of energy storage has the same energy.

SOLVE The molecule rotates end-over-end, as was shown in Figure 18.12. Because the molecule rotates about its midpoint, each of the nitrogen atoms undergoes uniform circular motion with radius $r = 0.06$ nm. We found in Chapter 7 that the speed of a particle in circular motion is $v = 2\pi r/T_{rot}$, where T_{rot} is the rotation period. Hence the rotational kinetic energy of the molecule is

$$\epsilon_{rot} = \frac{1}{2}mv^2 + \frac{1}{2}mv^2 = m\left(\frac{2\pi r}{T_{rot}}\right)^2 = \frac{4\pi^2 mr^2}{T_{rot}^2}$$

The energy associated with this mode is $\frac{1}{2}Nk_BT$, so the *average* rotational kinetic energy per molecule is

$$(\epsilon_{rot})_{avg} = \frac{1}{2}k_BT$$

Equating these two expressions for ϵ_{rot} gives us

$$\frac{4\pi^2 mr^2}{T_{rot}^2} = \frac{1}{2}k_BT$$

Thus the rotational period is

$$T_{rot} = \sqrt{\frac{8\pi^2 mr^2}{k_BT}} = 1.3 \times 10^{-12} \text{ s}$$

We evaluated T_{rot} at $T = 293$ K, using $m = 14$ u $= 2.34 \times 10^{-26}$ kg for each *atom*. The molecule's rotational frequency is

$$f = \frac{1}{T_{rot}} = 7.7 \times 10^{11} \text{ rev/s}$$

ASSESS This is a *very* high frequency, but these values are typical of molecular rotations.

STOP TO THINK 18.4 How many degrees of freedom does a bead on a fixed wire have?

   a. 1   b. 2   c. 3   d. 4   e. 5   f. 6

18.5 Thermal Interactions and Heat

We can now look in more detail at what happens when two systems at different temperatures interact with each other. Figure 18.14 shows a rigid, insulated container that is divided into two sections by a very thin membrane. The left side, which we'll call system 1, has N_1 atoms at an initial temperature T_{1i}. System 2 on the right has N_2 atoms at an initial temperature T_{2i}. The membrane is so thin that atoms can collide at the boundary as if the membrane were not there, yet it is a barrier that prevents atoms from moving from one side to the other. The situation is analogous, on an atomic scale, to basketballs colliding through a shower curtain.

Suppose that system 1 is initially at a higher temperature: $T_{1i} > T_{2i}$. This is not an equilibrium situation. The temperatures will change with time until they eventually reach a common final temperature T_f. If you *watch* the gases as one warms and the other cools, you see nothing happening. This interaction is quite different from a mechanical interaction in which, for example, you might see a piston move from one side toward the other. The only way in which the gases can interact is via molecular collisions at the boundary. This is a *thermal interaction,* and our goal is to understand how thermal interactions bring the systems to thermal equilibrium.

System 1 and system 2 begin with thermal energies

$$E_{1i} = \frac{3}{2}N_1 k_B T_{1i} = \frac{3}{2}n_1 R T_{1i}$$

$$E_{2i} = \frac{3}{2}N_2 k_B T_{2i} = \frac{3}{2}n_2 R T_{2i}$$

(18.38)

We've written the energies for monatomic gases; you could do the same calculation if one or both of the gases is diatomic by replacing the $\frac{3}{2}$ with $\frac{5}{2}$. Notice that we've omitted "th" from the subscript to keep the notation manageable.

The total energy of the combined systems is $E_{tot} = E_{1i} + E_{2i}$. As systems 1 and 2 interact, their individual thermal energies E_1 and E_2 can change but their sum E_{tot} remains constant. The system will have reached thermal equilibrium when the individual thermal energies reach final values E_{1f} and E_{2f} that no longer change.

The Systems Exchange Energy

Figure 18.15 shows a fast atom and a slow atom approaching the barrier from opposite sides. They undergo a perfectly elastic collision at the barrier. Although no net energy is lost in a perfectly elastic collision, the faster atom loses energy while the slower one gains energy. In other words, there is an energy *transfer* from the faster atom's side to the slower atom's side.

The average translational kinetic energy per molecule is directly proportional to the temperature: $\epsilon_{avg} = \frac{3}{2}k_B T$. Because $T_{1i} > T_{2i}$, the atoms in system 1 are, on average, more energetic than the atoms in system 2. Thus *on average* the collisions transfer energy from system 1 to system 2. But not in every collision. Sometimes a fast atom in system 2 collides with a slow atom in system 1, transferring energy from 2 to 1. But the net energy transfer, from all collisions, is from the warmer system 1 to the cooler system 2. In other words, **heat is the energy transferred *via collisions* between the more energetic (warmer) atoms on one side and the less energetic (cooler) atoms on the other.**

Insulation prevents heat from entering or leaving the container.

A thin barrier prevents atoms from moving from system 1 to 2 but still allows them to collide.

FIGURE 18.14 Two gases can interact thermally through a very thin barrier.

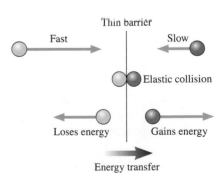

FIGURE 18.15 Collisions at the barrier transfer energy from faster molecules to slower molecules.

How do the systems "know" when they've reached thermal equilibrium? Energy transfer continues until the atoms on both sides of the barrier have the *same average translational kinetic energy*. Once the average translational kinetic energies are the same, there is no tendency for energy to flow in either direction. This is the state of thermal equilibrium, so the condition for thermal equilibrium is

$$(\epsilon_1)_{\text{avg}} = (\epsilon_2)_{\text{avg}} \qquad \text{(thermal equilibrium)} \tag{18.39}$$

where, as before, ϵ is the translational kinetic energy of an atom.

Because the average energies are directly proportional to the final temperatures, $\epsilon_{\text{avg}} = \frac{3}{2}k_{\text{B}}T_{\text{f}}$, thermal equilibrium is characterized by the macroscopic condition

$$T_{1\text{f}} = T_{2\text{f}} = T_{\text{f}} \qquad \text{(thermal equilibrium)} \tag{18.40}$$

In other words, **two thermally interacting systems reach a common final temperature *because* they exchange energy via collisions until the atoms on each side have, on average, equal translational kinetic energies.** This is a very important idea.

Equation 18.40 can be used to determine the equilibrium thermal energies. Because these are monatomic gases, $E_{\text{th}} = N\epsilon_{\text{avg}}$. Thus the equilibrium condition $(\epsilon_1)_{\text{avg}} = (\epsilon_2)_{\text{avg}} = (\epsilon_{\text{tot}})_{\text{avg}}$ implies

$$\frac{E_{1\text{f}}}{N_1} = \frac{E_{2\text{f}}}{N_2} = \frac{E_{\text{tot}}}{N_1 + N_2} \tag{18.41}$$

from which we can conclude

$$E_{1\text{f}} = \frac{N_1}{N_1 + N_2}E_{\text{tot}}$$
$$E_{2\text{f}} = \frac{N_2}{N_1 + N_2}E_{\text{tot}} \tag{18.42}$$

This result can also be written in terms of the number of moles. If we use $N = N_{\text{A}}n$ and note that the N_{A} cancels, Equation 18.42 becomes

$$E_{1\text{f}} = \frac{n_1}{n_1 + n_2}E_{\text{tot}}$$
$$E_{2\text{f}} = \frac{n_2}{n_1 + n_2}E_{\text{tot}} \tag{18.43}$$

Notice that $E_{1\text{f}} + E_{2\text{f}} = E_{\text{tot}}$, verifying that energy has been conserved even while being redistributed between the systems.

No work is done on either system, because the barrier has no macroscopic displacement, so the first law of thermodynamics is

$$Q_1 = \Delta E_1 = E_{1\text{f}} - E_{1\text{i}}$$
$$Q_2 = \Delta E_2 = E_{2\text{f}} - E_{2\text{i}} \tag{18.44}$$

As a homework problem you can show that $Q_1 = -Q_2$, as required by energy conservation. That is, the heat lost by one system is gained by the other. $|Q_1|$ is the quantity of heat that is transferred from the warmer gas to the cooler gas during the thermal interaction.

NOTE ▶ In general, the equilibrium thermal energies of the system are *not* equal. That is, $E_{1\text{f}} \neq E_{2\text{f}}$. They will be equal only if $N_1 = N_2$. Equilibrium is reached when the average translational kinetic energies in the two systems are equal; that is, when $(\epsilon_1)_{\text{avg}} = (\epsilon_2)_{\text{avg}}$, not when $E_{1\text{f}} = E_{2\text{f}}$. The distinction is important. Figure 18.16 summarizes these ideas. ◀

Collisions transfer energy from warmer system to cooler as more energetic atoms lose energy to less energetic atoms.

Thermal equilibrium occurs when the systems have the same average translational kinetic energy and thus the same temperature.

In general, the thermal energies $E_{1\text{f}}$ and $E_{2\text{f}}$ are *not equal.*

FIGURE 18.16 Equilibrium is reached when the atoms on each side have, on average, equal energies.

EXAMPLE 18.10 **A thermal interaction**
A sealed, insulated container has 2.0 g of helium at an initial temperature of 300 K on one side of a barrier and 10.0 g of argon at an initial temperature of 600 K on the other side.

a. How much heat energy is transferred, and in which direction?
b. What is the final temperature?

MODEL The systems start with different temperatures, so they are not in thermal equilibrium. Energy will be transferred via collisions from the argon to the helium until both systems have the same average molecular energy.

SOLVE

a. Let the helium be system 1. Helium has molar mass $M_{mol} = 4$ g/mol, so $n_1 = M/M_{mol} = 0.50$ mol. Similarly, argon has $M_{mol} = 40$ g/mol, so $n_2 = 0.25$ mol. The initial thermal energies of the two monatomic gases are

$$E_{1i} = \frac{3}{2}n_1 R T_{1i} = 225R = 1870 \text{ J}$$

$$E_{2i} = \frac{3}{2}n_2 R T_{2i} = 225R = 1870 \text{ J}$$

The systems start with *equal* thermal energies, but they are not in thermal equilibrium. The total energy is $E_{tot} = 3740$ J. In equilibrium, this energy is distributed between the two systems as

$$E_{1f} = \frac{n_1}{n_1 + n_2}E_{tot} = \frac{0.50}{0.75}3740 \text{ J} = 2493 \text{ J}$$

$$E_{2f} = \frac{n_2}{n_1 + n_2}E_{tot} = \frac{0.25}{0.75}3740 \text{ J} = 1247 \text{ J}$$

The heat entering or leaving each system is

$$Q_1 = Q_{He} = E_{1f} - E_{1i} = 623 \text{ J}$$

$$Q_2 = Q_{Ar} = E_{2f} - E_{2i} = -623 \text{ J}$$

The helium and the argon interact thermally via collisions at the boundary, causing 623 J of heat to be transferred from the warmer argon to the cooler helium.

b. These are constant-volume processes, thus $Q = nC_V\Delta T$. $C_V = \frac{3}{2}R$ for monatomic gases, thus the temperature changes are

$$\Delta T_{He} = \frac{Q_{He}}{\frac{3}{2}nR} = \frac{623 \text{ J}}{1.5(0.50 \text{ mol})(8.31 \text{ J/mol K})} = 100 \text{ K}$$

$$\Delta T_{Ar} = \frac{Q_{Ar}}{\frac{3}{2}nR} = \frac{-623 \text{ J}}{1.5(0.25 \text{ mol})(8.31 \text{ J/mol K})} = -200 \text{ K}$$

Both gases reach the common final temperature $T_f = 400$ K.

ASSESS $E_{1f} = E_{2f}$ because there are twice as many atoms in system 1.

The main idea of this section is that two systems reach a common final temperature not by magic or by a prearranged agreement but simply from the energy exchange of vast numbers of molecular collisions. Real interacting systems, of course, are separated by walls rather than our unrealistic thin membrane. As the systems interact, the energy is first transferred via collisions from system 1 into the wall and subsequently, as the cooler molecules collide with a warm wall, into system 2. That is, the energy transfer is $E_1 \rightarrow E_{wall} \rightarrow E_2$. This is still heat because the energy transfer is occurring via molecular collisions rather than mechanical motion.

STOP TO THINK 18.5 Systems A and B are interacting thermally. At this instant of time,

a. $T_A > T_B$.
b. $T_A = T_B$.
c. $T_A < T_B$.

A	B
$N = 1000$	$N = 2000$
$\epsilon_{avg} = 1.0 \times 10^{-20}$ J	$\epsilon_{avg} = 0.5 \times 10^{-20}$ J
$E_{th} = 1.0 \times 10^{-17}$ J	$E_{th} = 1.0 \times 10^{-17}$ J

18.6 Irreversible Processes and the Second Law of Thermodynamics

The previous section looked at the thermal interaction between a warm gas and a cold gas. Heat energy is transferred from the warm gas to the cold gas until they reach a common final temperature. But why isn't heat transferred from the cold gas to the warm gas, making the cold side colder and the warm side warmer? Such a process could still conserve energy, but it never happens. The transfer of heat energy from hot to cold is an example of an **irreversible process,** a process that can happen only in one direction.

Examples of irreversible processes abound. Stirring the cream in your coffee mixes the cream and coffee together. No amount of stirring ever unmixes them. If you shake a jar that has red marbles on the top and blue marbles on the bottom, the two colors are quickly mixed together. No amount of shaking ever separates them again. If you watched a movie of someone shaking a jar and saw the red and blue marbles separating, you would be certain that the movie was running backward. In fact, a reasonable definition of an irreversible process is one for which a backward-running movie shows a physically impossible process.

Figure 18.17a is a two-frame movie of a collision between two particles, perhaps two gas molecules. Suppose that sometime after the collision is over we could reach in and reverse the velocities of both particles. That is, replace vector $\vec{v}$ with vector $-\vec{v}$. Then, as in a movie playing backward, the collision would happen in reverse. This is the movie of Figure 18.17b.

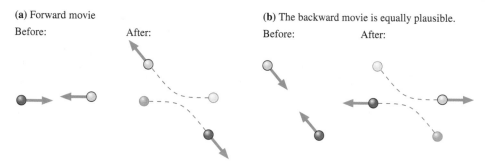

(a) Forward movie

Before: After:

(b) The backward movie is equally plausible.

Before: After:

FIGURE 18.17 Molecular collisions are reversible.

You cannot tell, just by looking at the two movies, which is really going forward and which is being played backward. Maybe Figure 18.17b was the original collision and Figure 18.17a is the backward version. Nothing in either collision looks wrong, and no measurements you might make on either would reveal any violations of Newton's laws. Interactions at the molecular level are reversible processes.

Contrast this with the two-frame car crash movies in Figure 18.18. Past and future are clearly distinct in an irreversible process, and the backward movie of Figure 18.18b is obviously wrong. But what has been violated in the backward movie? To have the crumpled car spring away from the wall would not violate any laws of physics we have so far discovered. It would simply require transforming the thermal energy of the car and wall back into the macroscopic center-of-mass energy of the car as a whole.

The paradox stems from our assertion that macroscopic phenomena can be understood on the basis of microscopic molecular motions. If the microscopic motions are all reversible, how can the macroscopic phenomena end up being irreversible? If reversible collisions can cause heat to be transferred from hot to cold, why do they never cause heat to be transferred from cold to hot? There must be another law of physics preventing it. The law we seek must, in some sense, be able to distinguish the past from the future.

(a) Forward movie

Before: After:

(b) The backward movie is physically impossible.

Before: After:

FIGURE 18.18 A car crash is irreversible.

Which Way to Equilibrium?

Stated another way, how do two systems initially at different temperatures "know" which way to go to reach equilibrium? Perhaps an analogy will help.

Figure 18.19 shows two boxes, numbered 1 and 2, containing identical balls. Box 1 starts with more balls than box 2, so $N_{1i} > N_{2i}$. Once every second, one ball is chosen at random and moved to the other box. This is a reversible process because a ball can move from box 2 to box 1 just as easily as from box 1 to box 2. What do you expect to see if you return several hours later?

Because balls are chosen at random, and because $N_{1i} > N_{2i}$, it's initially more likely that a ball will move from box 1 to box 2 than from box 2 to box 1. Sometimes a ball will move "backward" from box 2 to box 1, but overall there's a net movement of balls from box 1 to box 2. The system will evolve until $N_1 \approx N_2$. This is a stable situation—equilibrium!—with an equal number of balls moving in both directions.

But couldn't it go the other way, with N_1 getting even larger while N_2 decreases? In principle, any possible arrangement of the balls is possible in the same way that any number of heads are possible if you throw N coins in the air and let them fall. If you throw four coins, the odds are 1 in 2^4, or 1 in 16, of getting four heads. With four balls, the odds are 1 in 16 that, at a randomly chosen instant of time, you would find $N_1 = 4$. You wouldn't find that to be terribly surprising.

With 10 balls, the probability that $N_1 = 10$ is $0.5^{10} \approx 1/1000$. With 100 balls, the probability that $N_1 = 100$ has dropped to $\approx 10^{-30}$. With 10^{20} balls, the odds of finding all of them, or even most of them, in one box are so staggeringly small that it's safe to say it will "never" happen. Although each transfer is reversible, **the statistics of large numbers make it overwhelmingly more likely that the system will evolve toward a state in which $N_1 \approx N_2$ than toward a state in which $N_{1f} > N_{1i}$.**

The balls in our analogy represent energy. The total energy, like the total number of balls, is conserved, but molecular collisions can move energy between system 1 and system 2. Each collision is reversible, just as likely to transfer energy from 1 to 2 as from 2 to 1. But if $(\epsilon_{1i})_{avg} > (\epsilon_{2i})_{avg}$, and if we're dealing with two macroscopic systems where $N > 10^{20}$, then it's overwhelmingly likely that the net result of many, many collisions will be to transfer energy from system 1 to system 2 until $(\epsilon_{1f})_{avg} = (\epsilon_{2f})_{avg}$—in other words, for heat energy to be transferred from hot to cold.

The system reaches thermal equilibrium not by any plan or by outside intervention, but simply because **equilibrium is the *most probable* state in which to be.** It is *possible* that the system will move away from equilibrium, with heat moving from cold to hot, but remotely improbable in any realistic system. The consequence of a vast number of random events is that the system evolves in one direction, toward equilibrium, and not the other. **Reversible microscopic events lead to irreversible macroscopic behavior because some macroscopic states are vastly more probable than others.**

Order, Disorder, and Entropy

Figure 18.20 shows three different systems. At the top is a group of atoms arranged in a crystal-like lattice. This is a highly ordered and nonrandom system, with each atom's position precisely specified. Contrast this with the system on the bottom, where there is no order at all. The position of every atom was assigned entirely at random.

It is extremely improbable that the atoms in a container would *spontaneously* arrange themselves into the ordered pattern of the top picture. In a system of, say, 10^{20} atoms, the probability of this happening is similar to the probability that 10^{20} tossed coins will all be heads. We can safely say that it will never happen. By contrast, there are a vast number of arrangements like the one on the bottom that randomly fill the container.

The middle picture of Figure 18.20 is an in-between situation. This situation might arise as a solid melts. The positions of the atoms are clearly not completely

Balls are chosen at random and moved from one box to the other.

Box 1
N_1 balls

Box 2
N_2 balls

FIGURE 18.19 Two interacting systems. Balls are chosen at random and moved to the other box.

Increasing order
Decreasing entropy

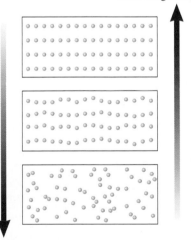

Increasing randomness
Increasing entropy

FIGURE 18.20 Ordered and disordered systems.

random, so the system preserves some degree of order. This in-between situation is more likely to occur spontaneously than the highly ordered lattice on the top, but is less likely to occur than the completely random system on the bottom.

Scientists and engineers use a state variable called **entropy** to measure the probability that a macroscopic state will occur. The ordered lattice, which has a very small probability of occurrence, has a very low entropy. The entropy of the randomly filled container is high. The entropy of the middle picture is somewhere in between. It is often said that entropy measures the amount of *disorder* in a system. The entropy in Figure 18.20 increases as you move from the ordered system on the top to the disordered system on the bottom.

Similarly, two thermally interacting systems with different temperatures have a low entropy. These systems are ordered in the sense that the faster atoms are on one side of the barrier, the slower atoms on the other. The most random possible distribution of energy, and hence the least ordered system, corresponds to the situation where the two systems are in thermal equilibrium with equal temperatures. Entropy increases as two systems with initially different temperatures move toward equilibrium. Entropy would decrease if heat energy moved from cold to hot, making the hot system hotter and the cold system colder.

Entropy can be calculated, but we'll leave that to more advanced courses. For our purposes, the *concept* of entropy as a measure of the disorder in a system, or of the probability that a macroscopic state will occur, is more important than a numerical value.

The Second Law of Thermodynamics

The fact that macroscopic systems evolve irreversibly toward equilibrium is a statement about nature that is not contained in any of the laws of physics we have encountered. It is, in fact, a new law of physics, one known as the **second law of thermodynamics.**

The formal statement of the second law of thermodynamics is given in terms of entropy:

> **Second law, formal statement** The entropy of an isolated system never decreases. The entropy either increases, until the system reaches equilibrium, or, if the system began in equilibrium, stays the same.

The qualifier "isolated" is most important. We can order the system by reaching in from the outside, perhaps using little tweezers to place the atoms in a lattice. Similarly, we can transfer heat from cold to hot by using a refrigerator. The second law is about what a system can or cannot do *spontaneously,* on its own, without outside intervention.

The second law of thermodynamics tells us that an isolated system evolves such that

- Order turns into disorder and randomness.
- Information is lost rather than gained.
- The system "runs down."

An isolated system never spontaneously generates order out of randomness. It is not that the system "knows" about order or randomness, but rather that there are vastly more states corresponding to randomness than there are corresponding to order. As collisions occur at the microscopic level, the laws of probability dictate that the system will, on average, move inexorably toward the most probable and thus most random macroscopic state.

The second law of thermodynamics is often stated in several equivalent but more informal versions. One of these, and the one most relevant to our discussion, is

Second law, informal statement #1 When two systems at different temperatures interact, heat energy is transferred spontaneously from the hotter to the colder system, never from the colder to the hotter.

The second law of thermodynamics is an independent statement about nature, separate from the first law. The first law is a precise statement about energy conservation. The second law, by contrast, is a *probabilistic* statement, based on the statistics of very large numbers. While it is conceivable that heat could spontaneously move from cold to hot, it will never occur in any realistic macroscopic system.

The irreversible evolution from less-likely macroscopic states to more-likely macroscopic states is what gives us a macroscopic direction of time. Stirring blends your coffee and cream, it never unmixes them. Friction causes an object to stop while increasing its thermal energy; the random atomic motions of thermal energy never spontaneously organize themselves into a macroscopic motion of the entire object. A plant in a sealed jar dies and decomposes to carbon and various gases; the gases and carbon never spontaneously assemble themselves into a flower. These are all examples of irreversible processes. They each show a clear direction of time, a distinct difference between past and future.

Thus another statement of the second law is

Second law, informal statement #2 The time direction in which the entropy of an isolated macroscopic system increases is "the future."

Establishing the "arrow of time" is one of the most profound implications of the second law of thermodynamics.

The second law of thermodynamics has important implications for issues ranging from how we as a society use energy and resources to biological evolution and the future of the universe. We'll return to some of these issues in the Summary to Part IV. In the meantime, the second law will be used in Chapter 19 to understand some of the practical aspects of the thermodynamics of engines.

STOP TO THINK 18.6   Two identical boxes each contain 1,000,000 molecules. In box A, 750,000 molecules happen to be in the left half of the box while 250,000 are in the right half. In box B, 499,900 molecules happen to be in the left half of the box while 500,100 are in the right half. At this instant of time,

a. The entropy of box A is larger than the entropy of box B.
b. The entropy of box A is equal to the entropy of box B.
c. The entropy of box A is smaller than the entropy of box B.

SUMMARY

The goal of Chapter 18 has been to understand the properties of a macroscopic system in terms of the microscopic behavior of its molecules.

GENERAL PRINCIPLES

Kinetic theory, the **micro/macro connection,** relates the macroscopic properties of a system to the motion and collisions of its atoms and molecules.

The Equipartition Theorem

Tells us how collisions distribute the energy in the system. The energy stored in each mode of the system (each **degree of freedom**) is $\frac{1}{2}Nk_BT$ or, in terms of moles, $\frac{1}{2}nRT$.

The Second Law of Thermodynamics

Tells us how collisions move a system toward equilibrium. The entropy of an isolated system can only increase or, in equilibrium, stay the same.

- Order turns into disorder and randomness.
- Systems run down.
- Heat energy is transferred spontaneously from the hotter to the colder system, never from colder to hotter.

IMPORTANT CONCEPTS

Pressure is due to the force of the molecules colliding with the walls.

$$p = \frac{1}{3}\frac{N}{V}mv_{rms}^2 = \frac{2}{3}\frac{N}{V}\epsilon_{avg}$$

The average translational kinetic energy of a molecule is $\epsilon_{avg} = \frac{3}{2}k_BT$. The temperature of the gas $T = \frac{2}{3k_B}\epsilon_{avg}$ measures the average translational kinetic energy.

Entropy measures the probability that a macroscopic state will occur or, equivalently, the amount of disorder in a system.

Increasing entropy ➡

The thermal energy of a system is
E_{th} = translational kinetic energy + rotational kinetic energy + vibrational energy

- **Monatomic gas** $E_{th} = \frac{3}{2}Nk_BT = \frac{3}{2}nRT$
- **Diatomic gas** $E_{th} = \frac{5}{2}Nk_BT = \frac{5}{2}nRT$
- **Elemental solid** $E_{th} = 3Nk_BT = 3nRT$

Heat is energy transferred via collisions from more-energetic molecules on one side to less-energetic molecules on the other. Equilibrium is reached when $(\epsilon_1)_{avg} = (\epsilon_2)_{avg}$, which implies $T_{1f} = T_{2f}$.

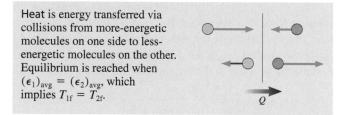

APPLICATIONS

The root-mean-square speed v_{rms} is the square root of the average of the squares of the molecular speeds:

$$v_{rms} = \sqrt{(v^2)_{avg}}$$

For molecules of mass m at temperature T,

$$v_{rms} = \sqrt{\frac{3k_BT}{m}}$$

Molar specific heats can be predicted from the thermal energy because $\Delta E_{th} = nC\Delta T$.

- **Monatomic gas** $C_V = \frac{3}{2}R$
- **Diatomic gas** $C_V = \frac{5}{2}R$
- **Elemental solid** $C = 3R$

TERMS AND NOTATION

kinetic theory	degrees of freedom	entropy
mean free path, λ	equipartition theorem	second law of thermodynamics
root-mean-square speed, v_{rms}	irreversible process	

EXERCISES AND PROBLEMS

Exercises

Section 18.1 Molecular Collisions

1. The number density of an ideal gas at STP is called the *Loschmidt number.* Calculate the Loschmidt number.
2. At what pressure will the mean free path in room-temperature (20°C) nitrogen be 1.0 m?
3. Integrated circuits are manufactured in vacuum chambers in which the air pressure is 1.0×10^{-10} mm of Hg. What are (a) the number density and (b) the mean free path of a molecule? Assume $T = 20$°C.
4. The mean free path of a molecule in a gas is 300 nm. What is the mean free path if the gas temperature is doubled at (a) constant volume and (b) constant pressure?
5. The pressure inside a tank of neon is 150 atm. The temperature is 25°C. On average, how many atomic diameters does a neon atom move between collisions?
6. A lottery machine uses blowing air to keep 2000 Ping-Pong balls bouncing around inside a 1.0 m × 1.0 m × 1.0 m box. The diameter of a Ping-Pong ball is 3.0 cm. What is the mean free path between collisions? Give your answer in cm.

Section 18.2 Pressure in a Gas

7. The molecules in a six-particle gas have velocities

$$\vec{v}_1 = (20\hat{\imath} + 30\hat{\jmath}) \text{ m/s} \qquad \vec{v}_4 = (60\hat{\imath} - 20\hat{\jmath}) \text{ m/s}$$
$$\vec{v}_2 = (-40\hat{\imath} + 70\hat{\jmath}) \text{ m/s} \qquad \vec{v}_5 = -50\hat{\jmath} \text{ m/s}$$
$$\vec{v}_3 = (-80\hat{\imath} - 10\hat{\jmath}) \text{ m/s} \qquad \vec{v}_6 = (40\hat{\imath} - 20\hat{\jmath}) \text{ m/s}$$

Calculate (a) $\vec{v}_{avg}$, (b) v_{avg}, and (c) v_{rms}.
8. Eleven molecules have speeds 15, 16, 17, . . . , 25 m/s. Calculate (a) v_{avg}, and (b) v_{rms}.
9. The number density in a container of argon gas is 2.0×10^{25} m^{-3}. The atoms are moving with an rms speed of 455 m/s. What are (a) the pressure and (b) the temperature inside the container?
10. 5.0×10^{23} nitrogen molecules collide with a 10 cm^2 wall each second. Assume that the molecules all travel with a speed of 400 m/s and strike the wall head on. What is the pressure on the wall?
11. Oxygen molecules collide with a 10 cm^2 wall. Assume that the molecules all travel with a speed of 500 m/s and strike the wall head on. How many collisions are there per second if the oxygen pressure is 1.0 atm?

12. At 100°C the rms speed of nitrogen molecules is 576 m/s. Nitrogen at 100°C and a pressure of 2.0 atm is held in a container with a 10 cm × 10 cm square wall. Estimate the rate of molecular collisions (collisions/s) on this wall.

Section 18.3 Temperature

13. What are the rms speeds of helium and argon atoms at 1000°C?
14. 30 m/s is a typical highway speed for a car. At what temperature do the molecules of nitrogen gas have an rms speed of 30 m/s?
15. A gas consists of a mixture of neon and argon. The rms speed of the neon atoms is 400 m/s. What is the rms speed of the argon atoms?
16. At what temperature do hydrogen molecules have the same rms speed as nitrogen molecules at 100°C?
17. At what temperature is the rms speed of oxygen molecules (a) half and (b) twice its value at STP?
18. By what factor does the rms speed of a molecule change if the temperature is increased from 20°C to 100°C?
19. At what temperature would the rms speed of hydrogen molecules be the speed of light (3.0×10^8 m/s)? There is no upper limit to temperature, but Einstein's theory of relativity says that no material particle can attain the speed of light. Consequently, our results for ϵ_{avg} and v_{rms} would need to be modified for very high temperatures and speeds.
20. Suppose you double the temperature of a gas at constant volume. Do the following change? And if so, by what factor?
   a. The average translational kinetic energy of a molecule.
   b. The rms speed of a molecule.
   c. The mean free path.
21. What is the total translational kinetic energy at STP of 1.0 mol of (a) hydrogen, (b) helium, and (c) oxygen?
22. During a physics experiment, helium gas is cooled to a temperature of 10 K at a pressure of 0.10 atm. What are (a) the mean free path in the gas, (b) the rms speed of the atoms, and (c) the average energy per atom?
23. The temperature at the center of the sun is $\approx 2 \times 10^7$ K. What are (a) the average kinetic energy and (b) the rms speed of a proton in the center of the sun?
24. The atmosphere of the sun consists mostly of hydrogen *atoms* (not molecules) at a temperature of 6000 K. What are (a) the average translational kinetic energy per atom and (b) the rms speed of the atoms?

Section 18.4 Thermal Energy and Specific Heat

25. A 6.0 m × 8.0 m × 3.0 m room contains air at 20°C. What is the room's thermal energy?

26. What is the thermal energy of 100 cm³ of lead at room temperature (20°C)?

27. The thermal energy of 1.0 mol of a substance is increased by 1.0 J. What is the temperature change if the system is (a) a monatomic gas, (b) a diatomic gas, and (c) a solid?

28. 1.0 mol of a monatomic gas interacts thermally with 1.0 mol of an elemental solid. The gas temperature decreases by 50°C at constant volume. What is the temperature change of the solid?

29. A rigid container holds 0.20 g of hydrogen gas. How much heat is needed to change the temperature of the gas
    a. From 50 K to 100 K?
    b. From 250 K to 300 K?
    c. From 550 K to 600 K?
    d. From 2250 K to 2300 K?

30. A cylinder of nitrogen gas has a volume of 15,000 cm³ and a pressure of 100 atm.
    a. What is the thermal energy of this gas at room temperature (20°C)?
    b. What is the mean free path in the gas?
    c. The valve is opened and the gas is allowed to expand slowly and isothermally until it reaches a pressure of 1.0 atm. What is the change in the thermal energy of the gas?

Section 18.5 Thermal Interactions and Heat

31. 2.0 mol of monatomic gas A initially has 5000 J of thermal energy. It interacts with 3.0 mol of monatomic gas B, which initially has 8000 J of thermal energy.
    a. Which gas has the higher initial temperature?
    b. What are the final thermal energies of each gas?

32. 4.0 mol of monatomic gas A initially has 9000 J of thermal energy. It interacts with 3.0 mol of monatomic gas B, which initially has 5000 J of thermal energy. How much heat energy is transferred between the systems, and in which direction, as they come to thermal equilibrium?

Problems

33. For a monatomic gas, what is the ratio of the volume per atom (V/N) to the volume *of* an atom when the mean free path equals the atomic diameter?

34. From what height must an oxygen molecule fall in a vacuum so that its kinetic energy at the bottom equals the average energy of an oxygen molecule at 300 K?

35. A gas at $p = 50$ kPa and $T = 300$ K has a mass density of 0.0802 kg/m³.
    a. Identify the gas.
    b. What is the rms speed of the atoms in this gas?
    c. What is the mean free path of the atoms in the gas?

36. Interstellar space, far from any stars, is filled with a very low density of hydrogen atoms (H, not H_2). The number density is about 1 atom/cm³ and the temperature is about 3 K.
    a. Estimate the pressure in interstellar space. Give your answer in Pa and in atm.

    b. What is the rms speed of the atoms?
    c. What is the edge length L of an $L \times L \times L$ cube of gas with 1.0 J of thermal energy?

37. Dust particles are ≈10 μm in diameter. They are pulverized rock, with $\rho \approx 2500$ kg/m³. If you treat dust as an ideal gas, what is the rms speed of a dust particle at 20°C?

38. Uranium has two naturally occurring isotopes. ^{238}U has a natural abundance of 99.3% and ^{235}U has an abundance of 0.7%. It is the rarer ^{235}U that is needed for nuclear reactors. The isotopes are separated by forming uranium hexafluoride UF_6, which is a gas, then allowing it to diffuse through a series of porous membranes. $^{235}UF_6$ has a slightly larger rms speed than $^{238}UF_6$ and diffuses slightly faster. Many repetitions of this procedure gradually separate the two isotopes. What is the ratio of the rms speed of $^{235}UF_6$ to that of $^{238}UF_6$?

39. Equation 18.3 is the mean free path of a particle through a gas of identical particles of equal radius. An electron can be thought of as a point particle with zero radius.
    a. Find an expression for the mean free path of an electron through a gas.
    b. Electrons travel 3 km through the Stanford Linear Accelerator (SLAC). In order for scattering losses to be negligible, the pressure inside the accelerator tube must be reduced to the point where the mean free path is at least 50 km. What is the maximum possible pressure inside the accelerator tube, assuming $T = 20$°C? Give your answer both in Pa and atm.

40. Figure P18.40 is a histogram showing the speeds of the molecules in a very small gas. What are (a) the most probable speed, (b) the average speed, and (c) the rms speed?

FIGURE P18.40

41. A 10-cm-diameter, 20-cm-long cylinder contains 2.0×10^{22} atoms of argon at a temperature of 50°C.
    a. What is the number density of the gas?
    b. What is the root-mean-square speed?
    c. What is $(v_x)_{rms}$, the rms value of the x-component of velocity?
    d. What is the rate at which atoms collide with one end of the cylinder?
    e. Determine the pressure in the cylinder using the results of kinetic theory.
    f. Determine the pressure in the cylinder using the ideal-gas law.

42. A 10 cm × 10 cm × 10 cm box contains 0.010 mol of nitrogen at 20°C. What is the rate of collisions (collisions/s) on one wall of the box?

43. A 100 cm³ box contains helium at a pressure of 2.0 atm and a temperature of 100°C. It is placed in thermal contact with a 200 cm³ box containing argon at a pressure of 4.0 atm and a temperature of 400°C.
    a. What is the initial thermal energy of each gas?
    b. What is the final thermal energy of each gas?

c. How much heat energy is transferred, and in which direction?

d. What is the final temperature?

e. What is the final pressure in each box?

44. 2.0 g of helium at an initial temperature of 300 K interacts thermally with 8.0 g of oxygen at an initial temperature of 600 K.

a. What is the initial thermal energy of each gas?

b. What is the final thermal energy of each gas?

c. How much heat energy is transferred, and in which direction?

d. What is the final temperature?

45. A gas of 1.0×10^{20} atoms or molecules has 1.0 J of thermal energy. Its molar specific heat at constant pressure is 20.8 J/mol K. What is the temperature of the gas?

46. How many degrees of freedom does a system have if $\gamma = 1.29$?

47. A system has f degrees of freedom. Show that $\gamma = (f + 2)/f$. Verify that this expression gives the values you learned in Chapter 17.

48. 1.0 mol of a monatomic gas and 1.0 mol of a diatomic gas are at 0°C. Both are heated at constant pressure until their volume doubles. What is the ratio $Q_{\text{diatomic}}/Q_{\text{monatomic}}$?

49. In the discussion following Equation 18.44 it was said that $Q_1 = -Q_2$. Prove that this is so.

50. A monatomic gas is adiabatically compressed to $\frac{1}{8}$ of its initial volume. Do each of the following quantities change? If so, does it increase or decrease, and by what factor? If not, why not?

a. The rms speed.

b. The mean free path.

c. The thermal energy of the gas.

d. The molar specific heat at constant volume.

51. Predict the molar specific heat at constant volume of (a) a two-dimensional monatomic gas and (b) a two-dimensional solid.

52. A diatomic molecule has only two rotational degrees of freedom because rotation about its own axis is "frozen out." This isn't a problem for polyatomic molecules with more than two atoms. They can rotate about any of three mutually perpendicular axes and thus have three rotational degrees of freedom.

a. Predict the molar specific heat at constant volume of a polyatomic molecule.

b. Water vapor has $\gamma = 1.31$. Is this consistent with your prediction?

53. Equal masses of hydrogen gas and oxygen gas are mixed together in a container and held at constant temperature. What is the hydrogen/oxygen ratio of (a) v_{rms}, (b) ϵ_{avg}, and (c) E_{th}?

54. A gas mixture consists of 2.0 mol of nitrogen and 1.0 mol of neon. What is the molar specific heat at constant volume of the mixture?

55. An experiment you're designing needs a gas with $\gamma = 1.50$. You recall from your physics class that no individual gas has this value, but it occurs to you that you could produce a gas with $\gamma = 1.50$ by mixing together a monatomic gas and a diatomic gas. What fraction of the molecules need to be monatomic?

56. n_1 moles of a monatomic gas and n_2 moles of a diatomic gas are mixed together in a container.

a. Derive an expression for the molar specific heat at constant volume of the mixture.

b. Show that your expression has the expected behavior if $n_1 \rightarrow 0$ or $n_2 \rightarrow 0$.

57. A 1.0 kg ball is at rest on the floor in a 2.0 m $\times$ 2.0 m $\times$ 2.0 m room of air at STP. Air is 80% nitrogen (N_2) and 20% oxygen (O_2) by volume.

a. What is the thermal energy of the air in the room?

b. What fraction of the thermal energy would have to be conveyed to the ball for it to be spontaneously launched to a height of 1.0 m?

c. By how much would the air temperature have to decrease to launch the ball?

d. Your answer to part c is so small as to be unnoticeable, yet this event never happens. Why not?

58. An inventor wants you to invest money with his company, offering you 10% of all future profits. He reminds you that the brakes on cars get extremely hot when they stop and that there is a large quantity of thermal energy in the brakes. He has invented a device, he tells you, that converts that thermal energy into the forward motion of the car. This device will take over from the engine after a stop and accelerate the car back up to its original speed, thereby saving a tremendous amount of gasoline. Now, you're a smart person, so he admits up front that this device is not 100% efficient, that there is some unavoidable heat loss to the air and to friction within the device, but the upcoming research for which he needs your investment will make those losses extremely small. You do also have to start the car with cold brakes after it has been parked awhile, so you'll still need a gasoline engine for that. Nonetheless, he tells you, his prototype car gets 500 miles to the gallon and he expects to be at well over 1000 miles to the gallon after the next phase of research. Should you invest? Base your answer on an analysis of the *physics* of the situation.

Challenge Problems

59. 1.0 mol of a diatomic gas with $C_V = \frac{5}{2}R$ has initial pressure p_i and volume V_i. The gas undergoes a process in which the pressure is directly proportional to the volume until the rms speed of the molecules has doubled.

a. Show this process on a pV diagram.

b. How much heat does this process require? Give your answer in terms of p_i and V_i.

60. At what temperature does the rms speed of (a) a nitrogen molecule and (b) a hydrogen molecule equal the escape speed from the earth's surface? (c) You'll find that these temperatures are very high, so you might think that the earth's gravity could easily contain both gases. But not all molecules move with v_{rms}. There is a distribution of speeds, and a small percentage of molecules have speeds several times v_{rms}. Bit by bit, a gas can slowly leak out of the atmosphere as its fastest molecules escape. A reasonable rule of thumb is that the earth's gravity can contain a gas only if the average translational kinetic energy per molecule is less than 1% of the kinetic energy needed to escape. Use this rule to show why the earth's atmosphere contains nitrogen but not hydrogen, even though hydrogen is the most abundant element in the universe.

61. Consider a container like that shown in Figure 18.14, with n_1 moles of a monatomic gas on one side and n_2 moles of a diatomic gas on the other. The monatomic gas has initial temperature T_{1i}. The diatomic gas has initial temperature T_{2i}.

    a. Show that the equilibrium thermal energies are

    $$E_{1f} = \frac{3n_1}{3n_1 + 5n_2}(E_{1i} + E_{2i})$$

    $$E_{2f} = \frac{5n_2}{3n_1 + 5n_2}(E_{1i} + E_{2i})$$

    b. Show that the equilibrium temperature is

    $$T_f = \frac{3n_1 T_{1i} + 5n_2 T_{2i}}{3n_1 + 5n_2}$$

    c. 2.0 g of helium at an initial temperature of 300 K interacts thermally with 8.0 g of oxygen at an initial temperature of 600 K. What is the final temperature? How much heat energy is transferred, and in which direction?

STOP TO THINK ANSWERS

Stop to Think 18.1: $\lambda_B > \lambda_A = \lambda_C > \lambda_D$. Increasing the volume makes the gas less dense, so λ increases. Increasing the radius makes the targets larger, so λ decreases. The mean free path doesn't depend on the atomic mass.

Stop to Think 18.2: c. Each v^2 increases by a factor of 16 but, after averaging, v_{rms} takes the square root.

Stop to Think 18.3: c. The average translational kinetic energy per molecule depends *only* on the temperature.

Stop to Think 18.4: b. The bead can slide along the wire (one degree of translational motion) and rotate around the wire (one degree of rotational motion).

Stop to Think 18.5: a. Temperature measures the average translational kinetic energy *per molecule,* not the thermal energy of the entire system.

Stop to Think 18.6: c. With 1,000,000 molecules, it's highly unlikely that 750,000 of them would spontaneously move into one side of the box. A state with a very small probability of occurrence has a very low entropy. Having an imbalance of only 100 out of 1,000,000 is well within what you might expect for random fluctuations. This is a highly probable situation and thus one of large entropy.

19 Heat Engines and Refrigerators

This coal-fired power plant is generating electricity by turning heat into work—but not very efficiently. Roughly two-thirds of the fuel's energy goes into the cooling water and is lost as waste heat.

▶ **Looking Ahead**

The goal of Chapter 19 is to investigate the physical principles that govern the operation of heat engines and refrigerators. In this chapter you will learn to:

■ Understand and analyze heat engines and refrigerators.

■ Understand the concept and significance of the Carnot engine.

■ Characterize the performance of a heat engine in terms of its thermal efficiency and that of a refrigerator in terms of its coefficient of performance.

■ Recognize that the second law of thermodynamics limits the efficiencies of heat engines and refrigerators.

◀ **Looking Back**

The material in this chapter depends on the first and second laws of thermodynamics. Most of the examples will be based on ideal gases. Please review:

■ Sections 16.5–16.6 Ideal gases.

■ Sections 17.2–17.4 Work, heat, and the first law of thermodynamics.

■ Section 18.6 The second law of thermodynamics.

The earliest humans learned to use the heat from fires to warm themselves and cook their food. They were transforming heat energy into thermal energy. But is there a way to transform heat into *work?* Can we use the energy released by the fuel to grind corn, pump water, accelerate cars, launch rockets, or do any other task in which a force is exerted through a distance?

The first practical device for turning heat into work was the steam engine. A steam engine boils water to make high-pressure steam, then uses the steam to push a piston and do work. The 19th and 20th centuries saw the development of the steam turbine, the gasoline engine, the jet engine, and other devices that transform the heat from burning fuel into useful work. These are the devices that power modern society.

"Heat engine" is the generic term for *any* device that uses a cyclical process to transform heat energy into work. The power plant shown in the photo and the engine in your car are examples of heat engines. A closely related concept is that of a *refrigerator,* a device that uses work to move heat energy from a cold object to a hot object. Our goal in this chapter is to investigate the physical principles that *all* heat engines and *all* refrigerators must obey. We'll discover that the second law of thermodynamics places sharp constraints on the maximum possible efficiency of heat engines and refrigerators.

A car engine transforms the chemical energy stored in the fuel into work and ultimately into the car's kinetic energy.

19.1 Turning Heat into Work

Thermodynamics is the branch of physics that studies the transformation of energy. Many practical devices are designed to transform energy from one form, such as the heat from burning fuel, into another, such as work. Chapters 17 and 18 established two laws of thermodynamics that any such device must obey:

First law Energy is conserved; that is, $\Delta E_{th} = W + Q$.

Second law Most macroscopic processes are irreversible. In particular, heat energy is transferred spontaneously from a hotter to a colder system but never from a colder system to a hotter system.

Our goal in this chapter is to discover what these two laws, especially the second law, imply about devices that turn heat into work. The questions we want to address include:

■ How does a practical device transform heat into work?
■ What are the limitations and restrictions on these energy transformations?

Much of this chapter will be an exercise in logical deduction. The reasoning is subtle but important.

Work Done by the System

In mechanics, we used the term "work" to mean the work done *on* the system by an external force; this definition considers work and heat to be separate but equivalent ways of transferring energy to the system. However, it is useful in practical thermodynamics to turn things around and speak of the work done on the environment *by* the system.

(a)

The external force does work $W = -F_{ext}\Delta x$.

Δx

$\vec{F}_{gas}$

$\vec{F}_{ext}$

The system does work $W_s = F_{gas}\Delta x$. $W_s = -W$ because $\vec{F}_{gas} = -\vec{F}_{ext}$.

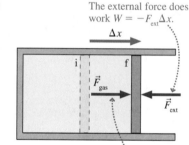

(b)

W_s is the area under the pV curve.

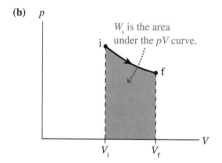

FIGURE 19.1 Forces $\vec{F}_{gas}$ and $\vec{F}_{ext}$ both do work as the piston moves.

Figure 19.1a shows a piston in a gas cylinder. The gas pressure pushes outward on the piston with force $\vec{F}_{gas}$. Some force in the environment, usually a force applied by a piston rod, pushes inward with force $\vec{F}_{ext}$. This external force is needed to keep the gas pressure from blowing the piston out of the cylinder. For any quasi-static process, where the system is essentially in equilibrium at all times, these two forces must balance: $\vec{F}_{gas} = -\vec{F}_{ext}$.

The work W done *on* the system is the work done by $\vec{F}_{ext}$ as the piston moves through a displacement Δx. You learned in Chapter 17 that W is the *negative* of the area under the pV curve of the process. But force $\vec{F}_{gas}$ also does work on the moving piston. Because $\vec{F}_{gas} = -\vec{F}_{ext}$, the work done by $\vec{F}_{gas}$, which we call the work W_s done *by* the system, has the same absolute value as the work W but the opposite sign. As Figure 19.1b shows, the work done *by* the system is

$$W_s = -W = \text{the area under the } pV \text{ curve} \qquad (19.1)$$

W_s is positive when energy is transferred *out* of the system by a mechanical interaction.

Work done by the environment and work done by the system are not mutually exclusive. Both $\vec{F}_{gas}$ and $\vec{F}_{ext}$ do work as the piston moves. Energy is transferred *into* the system as a gas is compressed, hence W is positive and W_s is negative. Energy is transferred *out* of the system as a gas expands, thus W is negative and W_s is positive.

NOTE ▶ When energy is transferred *into* a system, by compressing the gas, it is customary to say "the environment does work on the system." Similarly, when the gas pushes the piston out and transfers energy *out* of the system, we customarily say "the system does work on the environment." These expressions focus on the work that is positive. Neither is meant to imply that the "other" work isn't being done at the same time. ◀

The first law of thermodynamics $\Delta E_{th} = W + Q$ can be written in terms of W_s as

$$Q = W_s + \Delta E_{th} \qquad \text{(first law of thermodynamics)} \qquad (19.2)$$

It's easy to interpret this version of the first law. Because energy must be conserved, **any energy transferred into a system as heat is either used to do work or stored within the system as an increased thermal energy.**

Energy-Transfer Diagrams

Suppose you drop a hot rock into the ocean. Heat is transferred from the rock to the ocean until the rock and ocean are at the same temperature. Although the ocean warms up ever so slightly, ΔT_{ocean} is so small as to be completely insignificant. For all practical purposes, the ocean is infinite and unchangeable.

An **energy reservoir** is an object or a part of the environment so large that its temperature and thermal energy do not change when heat is transferred between the system and the reservoir. A reservoir at a higher temperature than the system is called a *hot reservoir*. A vigorously burning flame is a hot reservoir for small objects placed in the flame. A reservoir at a lower temperature than the system is called a *cold reservoir*. The ocean is a cold reservoir for the hot rock. We will use T_H and T_C to designate the temperatures of the hot and cold reservoirs.

Hot and cold reservoirs are idealizations, in the same category as frictionless surfaces and massless strings. No real object can maintain a perfectly constant temperature as heat is transferred in or out. Even so, an object can be accurately modeled as a reservoir if it is much larger than the system that thermally interacts with it.

Heat energy is transferred between a system and a reservoir if they have different temperatures. We will define

Q_H = amount of heat transferred to or from a hot reservoir.

Q_C = amount of heat transferred to or from a cold reservoir.

By definition, Q_H and Q_C are *positive* quantities. The direction of heat transfer, which determines the sign of Q in the first law, will always be clear as we deal with thermodynamic devices. For example, the heat transferred *from* the system to a cold reservoir is $Q = -Q_C$.

Figure 19.2a shows a heavy copper bar between a hot reservoir (at temperature T_H) and a cold reservoir (at temperature T_C). Heat Q_H is transferred from the hot reservoir into the copper and heat Q_C is transferred from the copper to the cold reservoir. Figure 19.2b is an **energy-transfer diagram** for this process. The hot reservoir is always drawn at the top, the cold reservoir at the bottom, and the system—the copper bar in this case—between them. The reservoirs and the system are connected by "pipes" that show the energy transfers. Figure 19.2b shows heat Q_H being transferred into the system and Q_C being transferred out.

These cooling towers are using evaporation to transfer heat energy to the atmosphere. The atmosphere is a cold reservoir.

(a) Heat is transferred from hot to cold.

Copper bar
Q_H Q_C
Fire   Ice
Hot reservoir at T_H Cold reservoir at T_C

(b) Heat energy is transferred from a hot reservoir to a cold reservoir. Energy conservation requires $Q_C = Q_H$.

Hot reservoir T_H
                    Q_H The copper bar is the system.
System
                    Q_C
Cold reservoir T_C

(c) The second law forbids a process in which heat is spontaneously transferred from a colder object to a hotter object.

Hot reservoir T_H
                    Q_H
System
                    Q_C
Cold reservoir T_C

FIGURE 19.2 Energy-transfer diagrams.

The first law of thermodynamics $Q = W_s + \Delta E_{th}$ refers to the *system*. Q is the net heat to the system. In this case, because Q_C is the quantity of heat *leaving* the system, $Q = Q_H - Q_C$. The copper bar does no work, so $W_s = 0$. The bar warms up when first placed between the two reservoirs, but it soon comes to a steady state where its temperature no longer changes. Then $\Delta E_{th} = 0$. Thus the first law tells us that $Q = Q_H - Q_C = 0$, from which we conclude that

$$Q_C = Q_H \qquad (19.3)$$

In other words, all of the heat transferred into the hot end of the rod is subsequently transferred out of the cold end. This isn't surprising. After all, we know that heat is transferred spontaneously from a hotter object to a colder object. Even so, there has to be some *means* by which the heat energy gets from the hotter object to the colder. The copper bar provides a route for the energy transfer, and $Q_C = Q_H$ is the statement that energy is conserved as it moves through the bar.

Contrast Figure 19.2b with Figure 19.2c. Figure 19.2c shows a system in which heat is being transferred from the cold reservoir to the hot reservoir. The first law of thermodynamics is not violated, because $Q_H = Q_C$, but the second law is. If there were such a system, it would allow the spontaneous (i.e., with no outside input or assistance) transfer of heat from a colder object to a hotter object. The process of Figure 19.2c is forbidden by the second law of thermodynamics.

Work into Heat and Heat into Work

Turning work into heat is easy. Take two rocks out of the ocean and rub them together vigorously until both are warmer. You're creating a mechanical interaction in which work increases the thermal energy of the rocks, or $W \rightarrow \Delta E_{th}$. Then toss both back into the ocean, where they return to their initial temperature as the excess thermal energy is transferred as heat from the slightly warmer rocks to the colder water ($\Delta E_{th} \rightarrow Q_C$). Figure 19.3 is the energy-transfer diagram for this process.

Hot reservoir T_H

W System

Q_C

Cold reservoir T_C

FIGURE 19.3 Work can be transformed into heat with 100% efficiency.

NOTE ▶ Energy-transfer diagrams show the "work pipe" entering or leaving the system from the side. ◀

The conversion of work into heat is 100% efficient. That is, *all* of the energy supplied to the system as work W is transferred into the ocean as heat Q_C. This perfect transformation of work into heat can continue as long as there is motion. (It was this continual production of heat energy in the boring of cannons that Count Rumford recognized as being in conflict with the caloric theory.)

But the reverse—transforming heat into work—isn't as easy.

Figure 19.4 shows the special gas cylinder that we used in Figure 17.13. Heat energy is transferred from the flame into the gas, but the temperature remains constant because the heat energy from the flame is used to do the work of lifting the mass. $\Delta E_{th} = 0$ in an isothermal expansion, so the first law is

$$W_s = Q \qquad (19.4)$$

The energy that is transferred into the gas as heat is transformed with 100% efficiency into work done by the gas as it lifts the mass. So why did we just say that transforming heat into work isn't as easy as transforming work into heat?

There's a difference. In Figure 19.3, where we transformed work into heat, the system *returned to its initial state*. We can repeat the process over and over, continuing to transform work into heat as long as there is motion. But Figure 19.4 is a one-time process. The gas does work once as it lifts the piston, but then the gas is no longer in its initial state. We cannot repeat the process. Extracting more and more work from the device of Figure 19.4 requires lifting the piston higher and higher until, ultimately, it reaches the end of the cylinder.

A one-time device for turning heat into work will eventually, when enough heat has been added, melt down or blow up! **To be practical, a device that transforms heat into work must return to its initial state at the end of the process**

Insulation

M

Piston

The gas does work by lifting the piston. Afterward, the gas is no longer in its initial state.

Δx

Gas

Flame

FIGURE 19.4 An isothermal process transforms heat into work, but only as a one-time event.

and be ready for continued use. You want your car engine to turn over and over as long as there is fuel.

Perhaps Figure 19.4 is just a bad idea for turning heat into work. Perhaps some other device can turn heat into work continuously. Interestingly, no one has ever invented a "perfect engine" that transforms heat into work with 100% efficiency *and returns to its initial state* so that it can continue to do work as long as there is fuel. Of course, that such a device has not been invented is not a proof that it can't be done. We'll provide a proof shortly, but for now we'll make the hypothesis that the process of Figure 19.5 is somehow forbidden.

Notice the asymmetry between Figures 19.3 and 19.5. The perfect transformation of work into heat is permitted, but the perfect transformation of heat into work is forbidden. This asymmetry parallels the asymmetry of the two processes in Figure 19.2. In fact, we'll soon see that the "perfect engine" of Figure 19.5 is forbidden for exactly the same reason: the second law of thermodynamics.

FIGURE 19.5 There are no perfect engines that turn heat into work with 100% efficiency.

19.2 Heat Engines and Refrigerators

We've alluded to the fact that your car engine operates in a *cycle* in which the temperature and pressure conditions inside the cylinders are repeated over and over. Another essential aspect of your car engine allows it to continue doing work. Your car engine has both a hot reservoir, the combustion of fuel, *and* a cold reservoir that consists of the radiator and the surrounding air. Heat is transferred *to* the engine when the spark plug ignites the fuel. Heat is transferred *out* of the engine to the cooling water and ultimately the air.

The steam generator at your local electric power plant works by boiling water to produce high-pressure steam that spins a turbine (which then spins a generator to produce electricity). The steam pressure and temperature drop as the steam goes through the turbine. The steam is then condensed to liquid water and pumped back to the boiler to start the process again. There are two crucial ideas here. First, the device works in a cycle, with the water returning to its initial conditions once a cycle. Second, heat is transferred to the water in the boiler, but heat is transferred *out* of the water in the condenser.

Car engines and steam generators are examples of what we call *heat engines*. A **heat engine** is any closed-cycle device that extracts heat Q_H from a hot reservoir, does useful work, and exhausts heat Q_C to a cold reservoir. A **closed-cycle device** is one that periodically *returns to its initial conditions,* repeating the same process over and over. That is, all state variables (pressure, temperature, thermal energy, and so on) return to their initial values once every cycle. Consequently, a heat engine can continue to do useful work for as long as it is attached to the reservoirs.

Figure 19.6 is the energy-transfer diagram of a heat engine. Unlike the forbidden "perfect engine" of Figure 19.5, a heat engine is connected both to a hot reservoir *and* a cold reservoir. You can think of a heat engine as "siphoning off" some of the heat that moves from the hot reservoir to the cold reservoir and transforming that heat into work—some heat, but not all.

Because the temperature and thermal energy of a heat engine return to their initial values at the end of each cycle, an important characteristic of *any* heat engine is that

$$(\Delta E_{th})_{net} = 0 \quad \text{(any heat engine, over one full cycle)} \quad (19.5)$$

Consequently, the first law of thermodynamics *for a full cycle* of a heat engine is $(\Delta E_{th})_{net} = Q + W = Q - W_s = 0$.

Let's define W_{out} to be the useful work done *by* the heat engine *per cycle*. The net heat transfer per cycle is $Q_{net} = Q_H - Q_C$, hence the first law applied to a heat engine is

$$W_{out} = Q_{net} = Q_H - Q_C \quad \text{(work per cycle done by a heat engine)} \quad (19.6)$$

The steam turbine in a modern power plant is an enormous device. Expanding steam does work by spinning the turbine.

1. Heat energy Q_H is transferred from the hot reservoir to the system.

2. Part of the energy is used to do useful work W_{out}.

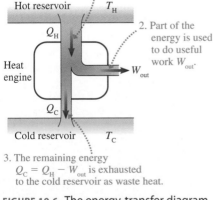

3. The remaining energy $Q_C = Q_H - W_{out}$ is exhausted to the cold reservoir as waste heat.

FIGURE 19.6 The energy-transfer diagram of a heat engine.

This is just energy conservation. The energy transferred into the engine (Q_H) and energy transferred out of the engine (Q_C and W_{out}) have to balance. The energy-transfer diagram of Figure 19.6 is a pictorial representation of Equation 19.6.

NOTE ▶ Equations 19.5 and 19.6 apply only to a *full cycle* of the heat engine. They are *not* valid for any of the individual processes that make up a cycle. ◀

For practical reasons, we would like an engine to do the maximum amount of work with the minimum amount of fuel. We can measure the performance of a heat engine in terms of its **thermal efficiency** η (lowercase Greek eta), defined as

$$\eta = \frac{W_{out}}{Q_H} = \frac{\text{what you get}}{\text{what you had to pay}} \qquad (19.7)$$

Using Equation 19.6 for W_{out}, we can also write the thermal efficiency as

$$\eta = 1 - \frac{Q_C}{Q_H} \qquad (19.8)$$

Figure 19.7 illustrates the idea of thermal efficiency.

A *perfect* heat engine would have $\eta_{perfect} = 1$. That is, it would be 100% efficient at converting heat from the hot reservoir (the burning fuel) into work. You can see from Equation 19.8 that a perfect engine would have no exhaust ($Q_C = 0$) and would not need a cold reservoir. Figure 19.5 has already suggested that there are no perfect heat engines, that an engine with $\eta = 1$ is impossible. A heat engine *must* exhaust energy to a cold reservoir. This energy transfer to the cold reservoir is called **waste heat.** It is energy that was extracted from the hot reservoir but *not* transformed to useful work.

Practical heat engines, such as car engines and steam generators, have thermal efficiencies in the range $\eta \approx 0.1–0.4$. This is not large. Well over half the input heat energy is "wasted." Can a clever designer do better, or is this some kind of physical limitation?

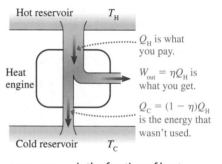

FIGURE 19.7 η is the fraction of heat energy that is transformed into useful work.

Q_H is what you pay.

$W_{out} = \eta Q_H$ is what you get.

$Q_C = (1 - \eta)Q_H$ is the energy that wasn't used.

STOP TO THINK 19.1 Rank in order, from largest to smallest, the work W_{out} performed by these four heat engines.

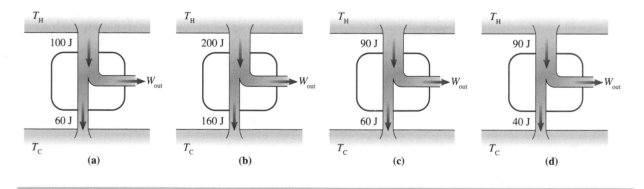

A Heat-Engine Example

This discussion has been somewhat abstract. To illustrate how these ideas actually work, Figure 19.8 shows a simple engine that converts heat into the work of lifting mass M. None of the state variables changes value as the mass is removed in step (c), but the piston must be locked in place to prevent the gas pressure from blowing it out once the mass is gone. The gas does work on the environment while lift-

FIGURE 19.8 A simple heat engine transforms heat into work.

ing the mass during step (b) (W_s is positive, W is negative). The environment does work on the gas during the compression of step (e) (W is positive, W_s is negative).

The net effect of this multistep process is to convert some of the fuel's energy into the useful work of lifting the mass. There has been no net change in the gas, which has returned to its initial pressure, volume, and temperature at the end of step (e). We can start the whole process over again and continue lifting masses (doing work) as long as we have fuel.

Figure 19.9 shows the heat-engine process on a pV diagram. It is a *closed cycle* because the gas returns to its initial conditions. No work is done during the isochoric process, so the *net* work done by the heat engine over the course of one full cycle is $W_{out} = (W_s)_{1\to2} + (W_s)_{3\to1}$.

Notice that the cyclical process of Figure 19.9 involves two *cooling processes* in which heat is transferred *from* the gas to the environment. Heat energy is transferred from hotter objects to colder objects, so the system *must* be connected to a cold reservoir with $T_C < T_{gas}$ during these two processes. A key to understanding heat engines is that they require both a heat source (burning fuel) *and* a heat sink (cooling water, the air, or something at a lower temperature than the system).

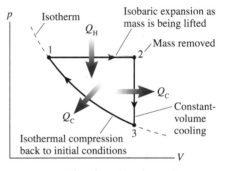

FIGURE 19.9 The closed-cycle pV diagram for the heat engine of Figure 19.8.

EXAMPLE 19.1 Analyzing a heat engine I
A heat engine follows the ideal-gas process shown in Figure 19.10. Analyze this engine to determine (a) the net work done per cycle, (b) the engine's thermal efficiency, and (c) the

FIGURE 19.10 The heat engine of Example 19.1.

engine's power output if it runs at 600 rpm. Assume the gas is monatomic.

MODEL The gas follows a closed cycle, starting and ending in state 1. The cycle consists of three distinct processes, each of which was studied in Chapters 16 and 17. For each of the three we need to determine the work done and the heat transferred.

SOLVE To begin, we can use the initial conditions at state 1 and the ideal-gas law to determine the number of moles of gas:

$$n = \frac{p_1V_1}{RT_1} = \frac{(200 \times 10^3 \, \text{Pa})(2.0 \times 10^{-4} \, \text{m}^3)}{(8.31 \, \text{J/mol K})(300 \, \text{K})} = 0.01604 \, \text{mol}$$

Process $1 \to 2$: The work done *on* the gas in the isobaric expansion is

$$W_{12} = -p\Delta V = -(200 \times 10^3 \, \text{Pa})(6.0 - 2.0) \times 10^{-4} \, \text{m}^3$$
$$= -80.0 \, \text{J}$$

Thus the work done *by* the gas is $(W_s)_{12} = +80.0$ J. We can use the ideal-gas law at constant pressure to find $T_2 = (V_2/V_1)T_1 = 3T_1 = 900$ K. The heat transfer during a constant-pressure process is

$$Q_{12} = nC_P\Delta T$$
$$= (0.01604 \text{ mol})(20.8 \text{ J/mol K})(900 \text{ K} - 300 \text{ K})$$
$$= 200.0 \text{ J}$$

where we used $C_P = \frac{5}{2}R$ for a monatomic ideal gas.

Process $2 \rightarrow 3$: No work is done in an isochoric process, so $(W_s)_{23} = 0$. The temperature drops back to 300 K, so the heat transfer is

$$Q_{23} = nC_V\Delta T$$
$$= (0.01604 \text{ mol})(12.5 \text{ J/mol K})(300 \text{ K} - 900 \text{ K})$$
$$= -120.0 \text{ J}$$

where we used $C_V = \frac{3}{2}R$.

Process $3 \rightarrow 4$: The gas returns to its initial state with volume V_1. The work done *on* the gas during an isothermal process is

$$W_{31} = -nRT \ln\left(\frac{V_1}{V_3}\right)$$
$$= -(0.01604 \text{ mol})(8.31 \text{ J/mol K})(300 \text{ K}) \ln\left(\frac{1}{3}\right)$$
$$= 43.9 \text{ J}$$

Thus $(W_s)_{31} = -43.9$ J. W_s is negative because the environment does work on the gas to compress it. An isothermal process has $\Delta E_{th} = 0$ and hence, from the first law,

$$Q_{31} = -W_{31} = -43.9 \text{ J}$$

Q is negative because the gas must be cooled as it is compressed to keep the temperature constant.

a. The *net* work done by the engine during one cycle is

$$W_{out} = (W_s)_{12} + (W_s)_{23} + (W_s)_{31} = 36.1 \text{ J}$$

As a consistency check, notice that the net heat transfer is

$$Q_{net} = Q_{12} + Q_{23} + Q_{31} = 36.1 \text{ J}$$

Equation 19.6 told us that a heat engine *must* have $W_{out} = Q_{net}$, and we see that it does.

b. The efficiency depends not on the net heat transfer but on the heat Q_H transferred into the engine from the flame. Heat is transferred in during process $1 \rightarrow 2$, where Q is positive, and out during processes $2 \rightarrow 3$ and $3 \rightarrow 1$, where Q is negative. Thus

$$Q_H = Q_{12} = 200.0 \text{ J}$$
$$Q_C = |Q_{23}| + |Q_{31}| = 163.9 \text{ J}$$

Notice that $Q_H - Q_C = 36.1$ J $= W_{out}$. In this heat engine, 200.0 J of heat from the hot reservoir does 36.1 J of useful work. Thus the thermal efficiency is

$$\eta = \frac{W_{out}}{Q_H} = \frac{36.1 \text{ J}}{200.0 \text{ J}} = 0.18 \text{ or } 18\%$$

This heat engine is far from being a perfect engine!

c. An engine running at 600 rpm goes through 10 cycles per second. The power output is the work done *per second*:

$$P_{out} = (\text{work per cycle}) \times (\text{cycles per second})$$
$$= 361 \text{ J/s} = 361 \text{ W}$$

ASSESS Although we didn't need Q_{net}, verifying that $Q_{net} = W_{out}$ was a check of self-consistency. Heat-engine analysis requires many calculations and offers many opportunities to get signs wrong. However, there are a sufficient number of self-consistency checks so that you can almost always spot calculational errors *if you check for them*. Also notice that we carried an extra significant figure in some of the intermediate calculations, such as $Q_H = 200.0$ J. Heat-engine calculations often require subtracting one number from another number of similar size; hence keeping an extra significant figure during intermediate calculations helps to minimize round-off error.

Let's think about this example a bit more before going on. We've said that a heat engine operates between a hot reservoir and a cold reservoir. Figure 19.10 doesn't explicitly show the reservoirs. Nonetheless, we know that heat is transferred from a hotter object to a colder object. Heat Q_H is transferred into the system during process $1 \rightarrow 2$ as the gas warms from 300 K to 900 K. For this to be true, the hot-reservoir temperature T_H must be ≥ 900 K. Likewise, heat Q_C is transferred from the system to the cold reservoir as the temperature drops from 900 K to 300 K in process $2 \rightarrow 3$. For this to be true, the cold-reservoir temperature T_C must be ≤ 300 K.

So we really don't know what the reservoirs are or their exact temperatures, but we can say with certainty that the hot-reservoir temperature T_H must exceed the highest temperature reached by the system and the cold-reservoir temperature T_C must be less than the coldest system temperature.

Refrigerators

Your house or apartment has a refrigerator. Very likely it has an air conditioner. The purpose of these devices is to make air that is cooler than its environment even colder. The first does so by blowing hot air out into a warm room, the second by blowing it out to the hot outdoors. You've probably felt the hot air exhausted by an air conditioner compressor or coming out from beneath the refrigerator.

At first glance, a refrigerator or air conditioner may seem to violate the second law of thermodynamics. After all, doesn't the second law forbid heat from being transferred from a colder object to a hotter object? Not quite: The second law says that heat is not *spontaneously* transferred from a colder to a hotter object. A refrigerator or air conditioner requires electric power to operate. They do cause heat to be transferred from cold to hot, but the transfer is "assisted" rather than spontaneous.

We'll define a **refrigerator** as any closed-cycle device that uses external work W_{in} to remove heat Q_C from a cold reservoir and exhaust heat Q_H to a hot reservoir. Figure 19.11 is the energy-transfer diagram of a refrigerator. The cold reservoir is the air inside the refrigerator or the air inside your house on a summer day. To keep the air cold, in the face of inevitable "heat leaks," the refrigerator or air conditioner compressor continuously removes heat from the cold reservoir and exhausts heat into the room or outdoors. You can think of a refrigerator as "pumping heat uphill," much as a water pump lifts water uphill.

Because a refrigerator, like a heat engine, is a cyclical device, $\Delta E_{th} = 0$. Conservation of energy requires

$$Q_H = Q_C + W_{in} \qquad (19.9)$$

To move energy from a colder to a hotter reservoir, a refrigerator must exhaust *more* heat to the outside than it removes from the inside. This has significant implications as to whether or not you can cool a room by leaving the refrigerator door open.

The thermal efficiency of a heat engine was defined as "what you get (useful work W_{out})" versus "what you pay for (fuel to supply Q_H)." By analogy, we define the **coefficient of performance** K of a refrigerator to be

$$K = \frac{Q_C}{W_{in}} = \frac{\text{what you get}}{\text{what you had to pay}} \qquad (19.10)$$

What you get, in this case, is the removal of heat from the cold reservoir. But you have to pay the electric company for the work needed to run the refrigerator. A better refrigerator will require less work to remove a given amount of heat, thus having a larger coefficient of performance.

A perfect refrigerator would require no work at all ($W_{in} = 0$) and would have $K_{perfect} = \infty$. But if Figure 19.11 had no work input, it would look just like Figure 19.2b. That device was forbidden by the second law of thermodynamics because, with no work input, heat would be able to move *spontaneously* from cold to hot.

We noted in Chapter 18 that the second law of thermodynamics can be stated several different but equivalent ways. We can now give a third statement:

> **Second law, informal statement #3** There are no perfect refrigerators with $K = \infty$.

Any real refrigerator or air conditioner *must* use work to move energy from the cold reservoir to the hot reservoir, hence $K < \infty$.

No Perfect Heat Engines

We hypothesized above that there are no perfect heat engines—that is, no heat engines like the one shown in Figure 19.5 with $Q_C = 0$ and $\eta = 1$. Now we're ready to prove this hypothesis. Figure 19.12 on the next page shows a hot reservoir at temperature T_H and a cold reservoir at temperature T_C. An ordinary refrigerator, one that obeys all the laws of physics, is operating between these two reservoirs.

This air conditioner transfers heat energy *from* the cool indoors *to* the hot exterior.

The amount of heat exhausted to the hot reservoir is larger than the amount of heat extracted from the cold reservoir.

External work is used to remove heat from a cold reservoir and exhaust heat to a hot reservoir.

FIGURE 19.11 The energy-transfer diagram of a refrigerator.

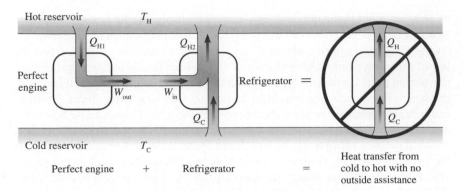

FIGURE 19.12 A perfect engine driving an ordinary refrigerator would be able to violate the second law of thermodynamics.

Suppose we had a perfect heat engine, one that takes in heat Q_H from the high-temperature reservoir and transforms that energy entirely into work W_{out}. If we had such a heat engine, we could use its output to provide the work input to the refrigerator. The two devices combined have no connection to the external world. That is, there's no net input or net output of work.

If we built a box around the heat engine and refrigerator, so that you couldn't see what was inside, the only thing you would observe is heat being transferred *with no outside assistance* from the cold reservoir to the hot reservoir. But a spontaneous or unassisted transfer of heat from a colder to a hotter object is exactly what the second law of thermodynamics forbids. Consequently, our assumption of a perfect heat engine must be wrong. Hence another statement of the second law of thermodynamics is:

> **Second law, informal statement #4** There are no perfect heat engines with $\eta = 1$.

Any real heat engine *must* exhaust waste heat Q_C to a cold reservoir.

Unanswered Questions

We noted that this chapter would be an exercise in logical deduction. By using only energy conservation and the fact that heat is not spontaneously transferred from cold to hot, we've been able to deduce that

- Heat engines and refrigerators exist.
- They must use a closed-cycle process, with $(\Delta E_{th})_{net} = 0$.
- There are no perfect heat engines. A heat engine *must* exhaust heat to a cold reservoir.
- There are no perfect refrigerators. A refrigerator *must* use external work.

This is a good start, but it leaves many unanswered questions. For example,

- With good design and good materials, can we make a heat engine whose thermal efficiency η approaches 1? Or is there an upper limit η_{max} that cannot be exceeded?
- If η has a maximum value, what is it?
- Likewise, is there an upper limit K_{max} for the coefficient of performance of a refrigerator? If so, what is it?

Section 19.6 will show that there is, indeed, an upper limit to η that no heat engine can exceed and an upper limit to K that no refrigerator can exceed. We'll be able to establish an actual value for η_{max} and will find that, for many practical engines, η_{max} is distressingly low.

STOP TO THINK 19.2 It's a really hot day and your air conditioner is broken. Your roommate says, "Let's open the refrigerator door and cool this place off." Will this work?

a. Yes.
b. No.
c. It might, but it will depend on how hot the room is.

19.3 Ideal-Gas Heat Engines

We will focus on heat engines that use a gas as the *working substance*. The gasoline engine in your car is an engine that alternately compresses and expands a gaseous fuel-air mixture. Diesel engines, jet engines, and gas turbines also fall into this category. The details of engines such as steam generators that rely upon phase changes will be deferred to more advanced courses.

A gas heat engine can be represented by a closed-cycle trajectory in the pV diagram, such as the one shown in Figure 19.13a. This observation leads to an interesting and important geometric interpretation of the work done by the system during one full cycle. You learned in Section 19.1 that the work done *by* the system is the area under the curve of a pV trajectory. As Figure 19.13b shows, the work done during a full cycle is the work W_{expand} done by the system as it expands from V_{min} to V_{max} plus the work $W_{compress}$ done by the system as it is compressed from V_{max} back to V_{min}. W_{expand} is positive and $W_{compress}$ is negative, thus

$$W_{out} = W_{expand} - |W_{compress}| = \text{area } inside \text{ the closed curve} \quad (19.11)$$

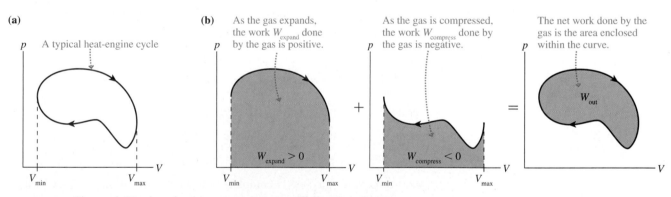

FIGURE 19.13 The work W_{out} done by the system during one full cycle is the area enclosed within the curve.

You can see that **the net work done by a gas heat engine during one full cycle is the area enclosed by the pV curve for the cycle.** A thermodynamic cycle with a larger enclosed area does more work than one with a smaller enclosed area. Notice that the gas must go around the pV trajectory in a *clockwise* direction for W_{out} to be positive. We'll see later that a refrigerator uses a counterclockwise (ccw) cycle.

Ideal-Gas Summary

We've learned a lot about ideal gases in the last three chapters. It will be helpful to summarize the major relationships we need in order to analyze heat engines. All gas processes obey the ideal-gas law $pV = nRT$ and the first law of thermodynamics $\Delta E_{th} = Q - W_s$. Results for specific gas processes are shown in Table 19.1 on the next page. This table shows W_s, the work done *by* the system, so the signs are opposite those in Chapter 17.

TABLE 19.1 Summary of ideal-gas processes

Process	Gas law	Work W_s	Heat Q	Thermal energy
Isochoric	$p_i/T_i = p_f/T_f$	0	$nC_V\Delta T$	$\Delta E_{th} = Q$
Isobaric	$V_i/T_i = V_f/T_f$	$p\Delta V$	$nC_P\Delta T$	$\Delta E_{th} = Q - W_s$
Isothermal	$p_iV_i = p_fV_f$	$nRT\ln(V_f/V_i)$ $pV\ln(V_f/V_i)$	$Q = W_s$	$\Delta E_{th} = 0$
Adiabatic	$p_iV_i^\gamma = p_fV_f^\gamma$ $T_iV_i^{\gamma-1} = T_fV_f^{\gamma-1}$	$(p_fV_f - p_iV_i)/(1-\gamma)$ $-nC_V\Delta T$	0	$\Delta E_{th} = -W_s$
Any	$p_iV_i/T_i = p_fV_f/T_f$	area under curve		$\Delta E_{th} = nC_V\Delta T$

TABLE 19.2 Properties of monatomic and diatomic gases

	Monatomic	Diatomic
E_{th}	$\frac{3}{2}nRT$	$\frac{5}{2}nRT$
C_V	$\frac{3}{2}R$	$\frac{5}{2}R$
C_P	$\frac{5}{2}R$	$\frac{7}{2}R$
γ	$\frac{5}{3} = 1.67$	$\frac{7}{5} = 1.40$

There is one entry in this table that you haven't seen before. The expression

$$W_s = \frac{p_fV_f - p_iV_i}{1 - \gamma} \quad \text{(work in an adiabatic process)} \quad (19.12)$$

for the work done in an adiabatic process follows from writing $W_s = -\Delta E_{th} = -nC_V\Delta T$, which you learned in Chapter 17, then using $\Delta T = \Delta(pV)/RT$ and the definition of γ. The proof will be left for a homework problem.

Finally, we learned in Chapter 18 that the thermal energy of an ideal gas depends only on its temperature. Table 19.2 summarizes the thermal energy, molar specific heats, and specific heat ratio $\gamma = C_P/C_V$ for monatomic and diatomic gases.

A Strategy for Heat-Engine Problems

8.12, 8.13 Activ Physics ONLINE

The engine of Example 19.1 was not a realistic heat engine, but it did illustrate the kinds of reasoning and computations involved in the analysis of a heat engine. A basic strategy for analyzing a heat engine is as follows.

(MP) PROBLEM-SOLVING STRATEGY 19.1 Heat-engine problems

MODEL Identify each process in the cycle.

VISUALIZE Draw the pV diagram of the cycle.

SOLVE There are several steps in the mathematical analysis.

- Use the ideal-gas law to complete your knowledge of n, p, V, and T at one point in the cycle.
- Use the ideal-gas law and equations for specific gas processes to determine p, V, and T at the beginning and end of each process.
- Calculate Q, W_s, and ΔE_{th} for each process.
- Find W_{out} by adding W_s for each process in the cycle. If the geometry is simple you can confirm this value by finding the area enclosed within the pV curve.
- Add just the *positive* values of Q to find Q_H.
- Verify that $(\Delta E_{th})_{net} = 0$. This is a self-consistency check to verify that you haven't made any mistakes.
- Calculate the thermal efficiency η and any other quantities you need to complete the solution.

ASSESS Is $(\Delta E_{th})_{net} = 0$? Do all the signs of W_s and Q make sense? Does η have a reasonable value? Have you answered the question?

EXAMPLE 19.2 **Analyzing a heat engine II**

A heat engine with a diatomic gas as the working substance uses the closed cycle shown in Figure 19.14. How much work does this engine do per cycle and what is its thermal efficiency?

FIGURE 19.14 The pV diagram for the heat engine of Example 19.2.

MODEL Processes $1 \rightarrow 2$ and $3 \rightarrow 4$ are isobaric. Processes $2 \rightarrow 3$ and $4 \rightarrow 1$ are isochoric.

VISUALIZE The pV diagram has been drawn for us in the problem statement.

SOLVE We know the pressure, volume, and temperature at state 4. The number of moles of gas in the heat engine is

$$n = \frac{p_4 V_4}{RT_4} = \frac{(101{,}300 \text{ Pa})(1.0 \text{ m}^3)}{(8.31 \text{ J/mol K})(300 \text{ K})} = 40.6 \text{ mol}$$

$p/T = $ constant during an isochoric process and $V/T = $ constant during an isobaric process. These allow us to find that $T_1 = T_3 = 900$ K and $T_2 = 2700$ K. This completes our knowledge of the state variables at all four corners of the diagram.

Process $1 \rightarrow 2$ is an isobaric expansion, so

$$(W_s)_{12} = p\Delta V = (3.0 \times 101{,}300 \text{ Pa})(2.0 \text{ m}^3) = 6.08 \times 10^5 \text{ J}$$

where we converted the pressure to pascals. The heat transfer during an isobaric expansion is

$$Q_{12} = nC_P\Delta T = (40.6 \text{ mol})(29.1 \text{ J/mol K})(1800 \text{ K})$$
$$= 21.27 \times 10^5 \text{ J}$$

where we used $C_P = \frac{7}{2}R$ for a diatomic gas. Then, using the first law,

$$\Delta E_{12} = Q_{12} - (W_s)_{12} = 15.19 \times 10^5 \text{ J}$$

Process $2 \rightarrow 3$ is an isochoric process, so $(W_s)_{23} = 0$ and

$$\Delta E_{23} = Q_{23} = nC_V\Delta T = -15.19 \times 10^5 \text{ J}$$

Notice that ΔT is *negative*.

Process $3 \rightarrow 4$ is an isobaric compression. Now ΔV is negative, so

$$(W_s)_{34} = p\Delta V = -2.03 \times 10^5 \text{ J}$$

and

$$Q_{34} = nC_P\Delta T = -7.09 \times 10^5 \text{ J}$$

Then $\Delta E_{th} = Q_{34} - (W_s)_{34} = -5.06 \times 10^5 \text{ J}$.

Process $4 \rightarrow 1$ is another constant-volume process, so again $(W_s)_{41} = 0$ and

$$\Delta E_{23} = Q_{23} = nC_V\Delta T = 5.06 \times 10^5 \text{ J}$$

The results of all four processes are shown in Table 19.3. The net results for W_{out}, Q_{net}, and $(\Delta E_{th})_{net}$ are found by summing the columns. As expected, $W_{out} = Q_{net}$ and $(\Delta E_{th})_{net} = 0$.

TABLE 19.3 Energy transfers in Example 19.2. All energies $\times 10^5$ J

Process	W_s	Q	ΔE_{th}
$1 \rightarrow 2$	6.08	21.27	15.19
$2 \rightarrow 3$	0	−15.19	−15.19
$3 \rightarrow 4$	−2.03	−7.09	−5.06
$4 \rightarrow 1$	0	5.06	5.06
Net	4.05	4.05	0

The work done during one cycle is $W_{out} = 4.05 \times 10^5$ J. Heat enters the system from the hot reservoir during processes $1 \rightarrow 2$ and $4 \rightarrow 1$, where Q is positive. Summing these gives $Q_H = 26.33 \times 10^5$ J. Thus the thermal efficiency of this engine is

$$\eta = \frac{W_{out}}{Q_H} = \frac{4.05 \times 10^5 \text{ J}}{26.33 \times 10^5 \text{ J}} = 0.15 = 15\%$$

ASSESS The verification that $W_{out} = Q_{net}$ and $(\Delta E_{th})_{net} = 0$ gives us great confidence that we didn't make any calculational errors. This engine may not seem very efficient, but η is quite typical of many real engines.

We noted in Example 19.1 that a heat engine's hot reservoir temperature T_H must exceed the highest temperature reached by the system and the cold reservoir temperature T_C must be less than the coldest system temperature. Although we don't know what the reservoirs are in Example 19.2, we can be sure that $T_H > 2700$ K and $T_C < 300$ K.

What is the thermal efficiency of this heat engine?

a. 0.10
b. 0.50
c. 0.25
d. 4
e. Can't tell without knowing Q_C.

The Brayton Cycle

The heat engines of Examples 19.1 and 19.2 have been educational but not realistic. As an example of a more realistic heat engine we'll look at the thermodynamic cycle known as the *Brayton cycle*. It is a reasonable model of a *gas turbine engine*. Gas turbines are used for electric power generation and as the basis for jet engines in aircraft and rockets. The *Otto cycle,* which describes the gasoline internal combustion engine, and the *diesel cycle,* which, not surprisingly, describes the diesel engine, will be the subject of homework problems.

Figure 19.15a is a schematic look at a gas turbine engine and Figure 19.15b is the corresponding pV diagram. To begin the Brayton cycle, air at an initial pressure p_1 is rapidly compressed in a *compressor.* This is an *adiabatic process,* with $Q = 0$, because there is no time for heat to be exchanged with the surroundings. Recall that an adiabatic compression raises the temperature of a gas by doing work on it, not by heating it, so the air leaving the compressor is very hot.

The hot gas flows into a combustion chamber. Fuel is continuously admitted to the combustion chamber where it mixes with the hot gas and is ignited, transferring heat to the gas at constant pressure and raising the gas temperature yet further. The high-pressure gas then expands, spinning a turbine that does some form of useful work. This is an adiabatic expansion, with $Q = 0$, that drops the temperature and pressure of the gas. Maximum work is produced with the maximum pressure drop, so the pressure at the end of the expansion through the turbine is back to p_1. However, the gas is still quite hot. The gas completes the cycle by flowing through a device called a **heat exchanger** that transfers heat energy to a cooling fluid. Large power plants are often sited on rivers or oceans in order to use the water for the cooling fluid in the heat exchanger.

This thermodynamic cycle, called a Brayton cycle, has two adiabatic processes—the compression and the expansion through the turbine—plus a constant-pressure heating and a constant-pressure cooling. There's no heat transfer during the adiabatic processes. The hot reservoir temperature must be $T_H \geq T_3$ in order for heat to be transferred into the gas during process $2 \rightarrow 3$. Similarly, the heat exchanger will remove heat from the gas only if $T_C \leq T_1$.

The thermal efficiency of any heat engine is

$$\eta = \frac{W_{out}}{Q_H} = 1 - \frac{Q_C}{Q_H}$$

Heat is transferred into the gas only during process $2 \rightarrow 3$. This is an isobaric process, so $Q_H = nC_P\Delta T = nC_P(T_3 - T_2)$. Similarly, heat is transferred out only during the isobaric process $4 \rightarrow 1$.

We have to be careful with signs. Q_{41} is negative, because the temperature decreases, but Q_C was defined as the *amount* of heat exchanged with the cold reservoir, a positive quantity. Thus

$$Q_C = |Q_{41}| = |nC_P(T_1 - T_4)| = nC_P(T_4 - T_1) \qquad (19.13)$$

A jet engine uses a modified Brayton cycle.

(a)

(b)

FIGURE 19.15 A gas turbine engine follows a Brayton cycle.

With these expressions for Q_H and Q_C, the thermal efficiency is

$$\eta_B = 1 - \frac{T_4 - T_1}{T_3 - T_2} \tag{19.14}$$

The subscript B indicates that this result applies to the Brayton cycle; it is not a general result. The efficiency depends on the temperatures, not the pressures or volumes. However, this expression isn't useful unless we compute all four temperatures. We can cast Equation 19.14 into a more useful form.

You learned in Chapter 17 that $pV^\gamma = $ constant during an adiabatic process, where $\gamma = C_p/C_V$ is the specific heat ratio. If we use $V = nRT/p$ from the ideal gas law, $V^\gamma = (nR)^\gamma T^\gamma p^{-\gamma}$. $(nR)^\gamma$ is a constant, so we can write $pV^\gamma = $ constant as

$$p^{(1-\gamma)}T^\gamma = \text{constant} \tag{19.15}$$

Equation 19.15 is a pressure-temperature relationship for an adiabatic process. Because $(T^\gamma)^{1/\gamma} = T$, we can simplify Equation 19.15 by raising both sides to the power $1/\gamma$. Doing so gives

$$p^{(1-\gamma)/\gamma}T = \text{constant} \tag{19.16}$$

during an adiabatic process.

Process $1 \rightarrow 2$ is an adiabatic process, hence

$$p_1^{(1-\gamma)/\gamma}T_1 = p_2^{(1-\gamma)/\gamma}T_2 \tag{19.17}$$

Isolating T_1 gives

$$T_1 = \frac{p_2^{(1-\gamma)/\gamma}}{p_1^{(1-\gamma)/\gamma}}T_2 = \left(\frac{p_2}{p_1}\right)^{(1-\gamma)/\gamma} T_2 = \left(\frac{p_{max}}{p_{min}}\right)^{(1-\gamma)/\gamma} T_2 \tag{19.18}$$

If we define the **pressure ratio** r_p as $r_p = p_{max}/p_{min}$, then T_1 and T_2 are related by

$$T_1 = r_p^{(1-\gamma)/\gamma}T_2 \tag{19.19}$$

The algebra of getting to Equation 19.19 was a bit tricky, but the final result is fairly simple.

Similarly, process $3 \rightarrow 4$ is an adiabatic process. The exact same reasoning leads to

$$T_4 = r_p^{(1-\gamma)/\gamma}T_3 \tag{19.20}$$

If we substitute these expressions for T_1 and T_4 into Equation 19.14, the efficiency is

$$\eta_B = 1 - \frac{T_4 - T_1}{T_3 - T_2} = 1 - \frac{r_p^{(1-\gamma)/\gamma}T_3 - r_p^{(1-\gamma)/\gamma}T_2}{T_3 - T_2} = 1 - \frac{r_p^{(1-\gamma)/\gamma}(T_3 - T_2)}{T_3 - T_2}$$

$$= 1 - r_p^{(1-\gamma)/\gamma}$$

Remarkably, all the temperatures cancel and we're left with an expression that depends only on the pressure ratio. Noting that $(1 - \gamma)$ is negative, it's useful to make one final change and write

$$\eta_B = 1 - \frac{1}{r_p^{(\gamma-1)/\gamma}} \tag{19.21}$$

Figure 19.16 is a graph of the efficiency of the Brayton cycle as a function of the pressure ratio, assuming $\gamma = 1.40$ for a diatomic gas such as air. You can see that η_B first grows quickly as the pressure ratio is increased, reaching $\approx 50\%$ at $r_p = 10$, then levels off. The pressure ratio is determined by the design of the compressor. A compressor with tighter-fitting parts and better seals will have a higher pressure ratio. However, any increase in efficiency beyond $\approx 50\%$ has to be weighed against the higher cost of a better compressor.

FIGURE 19.16 The efficiency of a Brayton cycle as a function of the pressure ratio r_p.

In Example 19.2 we found the thermal efficiency $\eta = W_{out}/Q_H$ by explicitly computing W_{out} and Q_H. Here, by contrast, we've determined the thermal efficiency of the Brayton cycle by using the relationship between the initial and final temperatures during an adiabatic process. The price we pay for this simplified analysis is that we didn't find an expression for the work done by a heat engine following the Brayton cycle. To calculate the work, which you can do as a homework problem, there's no avoiding the step-by-step analysis of the problem-solving strategy.

19.4 Ideal-Gas Refrigerators

Suppose that we were to operate a Brayton heat engine backward, going *ccw* in the *pV* diagram. Figure 19.17a shows a device for doing this. Figure 19.17b is its *pV* diagram and Figure 19.17c is the energy-transfer diagram. Starting from point 4, the gas is adiabatically compressed to increase its temperature and pressure. It then flows through a high-temperature heat exchanger where the gas *cools* at constant pressure from temperature T_3 to T_2. The gas then expands adiabatically, leaving it significantly colder at T_1 than it started at T_4. It completes the cycle by flowing through a low-temperature heat exchanger, where it *warms* back to its starting temperature.

Suppose that the low-temperature heat exchanger is a closed container of air surrounding a pipe through which the engine's cold gas is flowing. The heat-exchange process $1 \rightarrow 4$ *cools* the air in the container as it warms the gas flowing through the pipe. If you were to place eggs and milk inside this closed container, you would call it a refrigerator!

Going around a closed *pV* cycle in a ccw direction reverses the sign of *W* for each process in the cycle. Consequently, the area inside the curve of Figure 19.17b is W_{in}, the work done *on* the system. The work W_{in} done on the system is used to extract heat Q_C from the cold reservoir and exhaust a larger amount of heat $Q_H = Q_C + W_{in}$ to the hot reservoir. But where, in this situation, are the energy reservoirs?

Understanding a refrigerator is a little harder than understanding a heat engine. The key is to remember that **heat is always transferred from a hotter object to a colder object.** In particular,

■ The gas in a refrigerator can extract heat Q_C *from* the cold reservoir only if the gas temperature is *lower* than the cold-reservoir temperature T_C. Heat energy is then transferred *from* the cold reservoir *into* the colder gas.
■ The gas in a refrigerator can exhaust heat Q_H *to* the hot reservoir only if the gas temperature is *higher* than the hot-reservoir temperature T_H. Heat energy is then transferred *from* the warmer gas *into* the hot reservoir.

These two requirements place severe constraints on the thermodynamics of a refrigerator. Because there is no reservoir colder than T_C, the gas cannot reach a temperature lower than T_C by heat exchange. The gas in a refrigerator *must* use an adiabatic expansion ($Q = 0$) to lower the temperature below T_C. Likewise, a gas refrigerator requires an adiabatic compression to raise the gas temperature above T_H.

The reversed Brayton cycle of Figure 19.17b does, indeed, have two adiabatic processes. The adiabatic expansion lowers the temperature to T_1, then heat Q_C is transferred *from* the cold reservoir *to* the gas during process $1 \rightarrow 4$. Consequently, the cold reservoir temperature must be $T_C \geq T_4$. Contrast this with the same cycle run clockwise (cw) as a heat engine, where we saw that the cold reservoir must be $T_C \leq T_1$.

Similar reasoning applies on the hot side. In order for heat Q_H to be transferred *into* the hot reservoir during process $3 \rightarrow 2$, the hot reservoir temperature must be

(a)

(b)

(c)

FIGURE 19.17 A refrigerator that extracts heat from the cold reservoir and exhausts heat to the hot reservoir.

$T_H \leq T_2$. This requirement of the high-temperature reservoir differs distinctly from the Brayton heat engine, which required $T_H \geq T_3$. Figure 19.18 compares a Brayton-cycle heat engine to a Brayton-cycle refrigerator.

These cooling coils are the refrigerator's high-temperature heat exchanger. Heat energy is being transferred from hot gas inside the coils to the cooler room air.

FIGURE 19.18 A comparison of a Brayton-cycle heat engine to a Brayton-cycle refrigerator.

The important point—a point that will be important in the next section—is that a Brayton refrigerator is *not* simply a Brayton heat engine running backward. To make a Brayton refrigerator you must both reverse the cycle *and* change the hot and cold reservoirs.

NOTE ▶ Some heat engines cannot be converted to refrigerators under any circumstances. We'll leave it as a homework problem to show that the heat engine of Example 19.2, if run backward, is a total loser. Its energy-transfer diagram, shown in Figure 19.19, shows work being done to transfer energy "downhill" even faster than it would move spontaneously from hot to cold! ◀

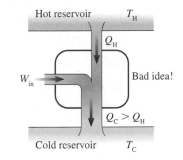

FIGURE 19.19 This is the energy-transfer diagram if the heat engine of Example 19.2 is run backward.

EXAMPLE 19.3 **Analyzing a refrigerator**

A refrigerator using helium gas operates on a reversed Brayton cycle with a pressure ratio of 5.0. Prior to compression, the gas occupies 100 cm³ at a pressure of 150 kPa and a temperature of −23°C. Its volume at the end of the expansion is 80 cm³. What are the refrigerator's coefficient of performance and its power input if it operates at 60 cycles per second?

MODEL The Brayton cycle has two adiabatic processes and two isobaric processes. The work per cycle needed to run the refrigerator is $W_{in} = Q_H - Q_C$, hence we can determine both the coefficient of performance and the power requirements from Q_H and Q_C. Heat energy is transferred only during the two isobaric processes.

VISUALIZE Figure 19.20 shows the pV cycle. We know from the pressure ratio of 5.0 that the maximum pressure is 750 kPa. Neither V_2 nor V_3 is known.

SOLVE To calculate heat we're going to need the temperatures at the four corners of the cycle. First, we can use the conditions of state 4 to find the number of moles of helium:

$$n = \frac{p_4 V_4}{R T_4} = 0.00722 \text{ mol}$$

FIGURE 19.20 A Brayton-cycle refrigerator.

Process $1 \rightarrow 4$ is isobaric, hence temperature T_1 is

$$T_1 = \frac{V_1}{V_4} T_4 = (0.80)(250 \text{ K}) = 200 \text{ K} = -73°C$$

With Equation 19.16 we found that the quantity $p^{(1-\gamma)/\gamma} T$ remains constant during an isobaric process. Helium is a monatomic gas with $\gamma = \frac{5}{3}$, so $(1 - \gamma)/\gamma = -\frac{2}{5} = -0.40$. For the adiabatic compression $4 \rightarrow 3$,

$$p_3^{-0.40}T_3 = p_4^{-0.40}T_4$$

Solving for T_3 gives

$$T_3 = \left(\frac{p_4}{p_3}\right)^{-0.40} T_4 = \left(\frac{1}{5}\right)^{-0.40}(250 \text{ K}) = 476 \text{ K} = 203°C$$

The same analysis applied to the $2 \rightarrow 1$ adiabatic expansion gives

$$T_2 = \left(\frac{p_1}{p_2}\right)^{-0.40} T_1 = \left(\frac{1}{5}\right)^{-0.40}(200 \text{ K}) = 381 \text{ K} = 108°C$$

Now we can use $C_P = \frac{5}{2}R = 20.8 \text{ J/molK}$ for a monatomic gas to compute the heat transfers:

$$Q_H = |Q_{32}| = nC_P(T_3 - T_2)$$

$$= (0.00722 \text{ mol})(20.8 \text{ J/mol K})(95 \text{ K}) = 14.27 \text{ J}$$

$$Q_C = |Q_{14}| = nC_P(T_4 - T_1)$$

$$= (0.00722 \text{ mol})(20.8 \text{ J/mol K})(50 \text{ K}) = 7.51 \text{ J}$$

Thus the work *input* to the refrigerator is $W_{in} = Q_H - Q_C = 6.76$ J. During each cycle, 6.76 J of work are done *on* the gas to extract 7.51 J of heat from the cold reservoir. 14.27 J of heat are then exhausted into the hot reservoir.

The refrigerator's coefficient of performance is

$$K = \frac{Q_C}{W_{in}} = \frac{7.51 \text{ J}}{6.76 \text{ J}} = 1.1$$

The power input needed to run the refrigerator is

$$P_{in} = 6.76\frac{\text{J}}{\text{cycle}} \times 60\frac{\text{cycles}}{\text{s}} = 406\frac{\text{J}}{\text{s}} = 406 \text{ W}$$

ASSESS These are fairly realistic values for a kitchen refrigerator. You pay your electric company for providing the work W_{in} that operates the refrigerator. The cold reservoir is the freezer compartment. The cold temperature T_C must be higher than T_4 ($T_C > -23°C$) in order for heat to be transferred *from* the cold reservoir *to* the gas. A typical freezer temperature is $-15°C$, so this condition is satisfied. The hot reservoir is the air in the room. The back and underside of a refrigerator have heat-exchanger coils where the hot gas, after compression, transfers heat to the air. The hot temperature T_H must be less than T_2 ($T_H < 108°C$) in order for heat to be transferred *from* the gas *to* the air. An air temperature $\approx 25°C$ under a refrigerator satisfies this condition.

STOP TO THINK 19.4 What, if anything, is wrong with this refrigerator?

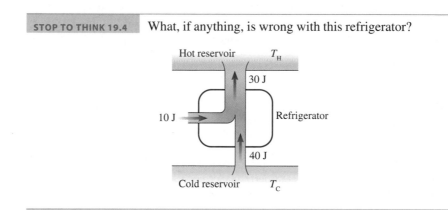

19.5 The Limits of Efficiency

Thermodynamics has its historical roots in the development of the steam engine and other machines of the early industrial revolution. Early steam engines, built on the basis of experience rather than scientific understanding, were not very efficient at converting fuel energy into work. The first major theoretical analysis of heat engines was published by the French engineer Sadi Carnot in 1824. Interestingly, his analysis *precedes* Joule's discovery of the full statement of conservation of energy. The question that Carnot raised was one we posed at the end of Section 19.3: Can we make a heat engine whose thermal efficiency η approaches 1, or is there an upper limit η_{max} that cannot be exceeded? To frame the question more clearly, imagine that we have a hot reservoir at temperature T_H and a cold reservoir at T_C. What is the most efficient heat engine (maximum η) that can operate between these two energy reservoirs? Similarly, what is the most efficient refrigerator (maximum K) that can operate between the two reservoirs?

We just saw that a refrigerator is, in some sense, a heat engine running backward. We might thus suspect that the most efficient heat engine may be related to the most efficient refrigerator. Suppose we have a heat engine that we can turn into a refrigerator by reversing the direction of operation, thus changing the direction of the energy transfers, and with *no other changes*. In particular, the heat engine and the refrigerator operate between the same two energy reservoirs at temperatures T_H and T_C.

> **NOTE** ▶ The heat engine we're looking for cannot be a Brayton-cycle heat engine. As we found, a Brayton-cycle refrigerator requires reversing the heat engine's direction of operation *and* changing the temperatures of the energy reservoirs. ◀

Figure 19.21a shows such a heat engine and its corresponding refrigerator. Notice that the refrigerator has *exactly the same* work and heat transfer as the heat engine, only in the opposite directions. A device that can be operated as either a heat engine or a refrigerator between the same two energy reservoirs and with the same energy transfers, with only their direction changed, is called a **perfectly reversible engine.** A perfectly reversible engine is an idealization, as was the concept of a perfectly elastic collision. Nonetheless, it will allow us to establish limits that no real engine can exceed.

FIGURE 19.21 If a perfectly reversible heat engine is used to operate a perfectly reversible refrigerator, the two devices exactly cancel each other. Consequently no net work is done and no net heat is transferred.

Suppose we have a perfectly reversible heat engine and a perfectly reversible refrigerator (the same device running backward) operating between a hot reservoir at temperature T_H and a cold reservoir at temperature T_C. Because the work W_{in} needed to operate the refrigerator is exactly the same as the useful work W_{out} done by the heat engine, we can use the heat engine, as shown in Figure 19.21b, to drive the refrigerator. The heat Q_C that the engine exhausts to the cold reservoir is exactly the same as the heat Q_C that the refrigerator extracts from the cold reservoir. Similarly, the heat Q_H that the engine extracts from the hot reservoir matches the heat Q_H that the refrigerator exhausts to the hot reservoir. Consequently, there is no net heat transfer in either direction. The refrigerator exactly replaces all the heat energy that had been transferred out of the hot reservoir by the heat engine.

You may want to compare the reasoning used here with the reasoning we used with Figure 19.12. There we tried to use the output of a "perfect" heat engine to run a refrigerator but did *not* succeed.

A Perfectly Reversible Engine Has Maximum Efficiency

Now we've arrived at the critical step in the reasoning. Suppose I claim to have a heat engine that can operate between temperatures T_H and T_C with *more* efficiency than a perfectly reversible engine. Figure 19.22 shows the output of this heat engine operating the same perfectly reversible refrigerator that we used in Figure 19.21b.

FIGURE 19.22 A heat engine more efficient than a perfectly reversible engine could be used to violate the second law of thermodynamics.

Recall that the thermal efficiency and the work of a heat engine are

$$\eta = \frac{W_{out}}{Q_H} \quad \text{and} \quad W_{out} = Q_H - Q_C$$

If the new heat engine is more efficient than the perfectly reversible engine it replaces, it needs *less* heat Q_H from the hot reservoir to perform the *same* work W_{out}. If Q_H is less while W_{out} is the same, then Q_C must also be less. That is, the new heat engine exhausts less heat to the cold reservoir than does the perfectly reversible heat engine.

When this new heat engine drives the perfectly reversible refrigerator, the heat it exhausts to the cold reservoir is *less* than the heat extracted from the cold reservoir by the refrigerator. Similarly, this engine extracts *less* heat from the hot reservoir than the refrigerator exhausts. Thus the net result of using this heat engine to operate a perfectly reversible refrigerator is that heat is transferred from the cold reservoir to the hot reservoir *without outside assistance.*

But this can't happen. It would violate the second law of thermodynamics. Hence we have to conclude that no heat engine operating between reservoirs at temperatures T_H and T_C can be more efficient than a perfectly reversible engine. This very important conclusion is another version of the second law:

> **Second law, informal statement #5** No heat engine operating between reservoirs at temperatures T_H and T_C can be more efficient than a perfectly reversible engine operating between these temperatures.

The answer to our question "Is there a maximum η that cannot be exceeded?" is a clear "Yes!" The maximum possible efficiency η_{max} is that of a perfectly reversible engine. Because the perfectly reversible engine is an idealization, any real engine will have an efficiency less than η_{max}.

A similar argument shows that no refrigerator can be more efficient than a perfectly reversible refrigerator. If we had such a refrigerator, and if we ran it with the output of a perfectly reversible heat engine, we could transfer heat from cold to hot with no outside assistance. Thus:

Second law, informal statement #6 No refrigerator operating between reservoirs at temperatures T_H and T_C can have a coefficient of performance larger than that of a perfectly reversible refrigerator operating between these temperatures.

Conditions for a Perfectly Reversible Engine

This argument tells us that η_{max} and K_{max} exist, but it doesn't tell us what they are. Our final task will be to "design" and analyze a perfectly reversible engine. Under what conditions is an engine reversible?

An engine transfers energy by both mechanical and thermal interactions. Mechanical interactions are pushes and pulls. The environment does work on the system, transferring energy into the system by pushing in on a piston. The system transfers energy back to the environment by pushing out on the piston. This exchange of energy is perfectly reversible if two conditions are met:

1. The energy transferred out (W_{out}) equals the energy transferred in (W_{in}), and
2. The system returns to its initial state, with no change of temperature or pressure.

Both these conditions will be met only if the motion is *frictionless*. The slightest bit of friction will prevent the mechanical transfer of energy from being perfectly reversible.

The circumstances under which heat transfer can be *completely* reversed aren't quite so obvious. After all, Chapter 18 emphasized the *irreversible* nature of heat transfer. If objects A and B are in thermal contact, with $T_A > T_B$, then heat energy is transferred from A to B. But the second law of thermodynamics prohibits a heat transfer from B back to A. Likewise, heat can move from B to A, but not from A to B, if $T_A < T_B$. Heat transfer through a temperature *difference* is an irreversible process.

But suppose $T_A = T_B$. With no temperature difference, any heat that is transferred from A to B can, at a later time, be transferred from B back to A. This transfer wouldn't violate the second law, which prohibits only heat transfer from a colder object to a hotter object. Now you might object, and rightly so, that heat *can't* move from A to B if they are at the same temperature because heat, by definition, is the energy transferred between two objects at different temperatures.

This is true, so let's consider a limiting case in which $T_A = T_B + dT$. That is, the temperature difference is infinitesimal. Heat is transferred from A to B, but *very slowly!* If you later try to make the heat move from B back to A, the second law will prevent you from doing so with perfect precision. But because the temperature difference is infinitesimal, you'll be missing only an infinitesimal amount dQ of heat. You can transfer heat reversibly in the limit $dT \rightarrow 0$, but you must be prepared to spend an infinite amount of time doing so.

Thus the thermal transfer of energy is reversible if the heat is transferred infinitely slowly in an isothermal process. This is an idealization, but so are completely frictionless processes. Nonetheless, we can now say that a perfectly reversible engine must use only two types of processes:

1. Frictionless mechanical interactions with no heat transfer ($Q = 0$), and
2. Thermal interactions in which heat is transferred in an isothermal process ($\Delta E_{th} = 0$).

Any engine that uses only these two types of processes is called a **Carnot engine.** A Carnot engine is a perfectly reversible engine; thus it has the maximum possible thermal efficiency η_{max} and, if operated as a refrigerator, the maximum possible coefficient of performance K_{max}.

19.6 The Carnot Cycle

8.14   Activ Physics

No real engine is perfectly reversible, so a Carnot engine is an idealization. Nonetheless, an analysis of the Carnot engine will allow us to establish a maximum possible thermal efficiency that no real heat engine can exceed.

The definition of a Carnot engine does not specify whether the engine's working substance is a gas or a liquid. It makes no difference. Our argument that a perfectly reversible engine is the most efficient possible heat engine depended only on the engine's reversibility. It did not depend on any details of how the engine is constructed or what it uses for a working substance. Consequently, **any Carnot engine operating between T_H and T_C must have exactly the same efficiency as any other Carnot engine operating between the same two energy reservoirs.** If we can determine the thermal efficiency of one Carnot engine, we'll know the efficiency of all Carnot engines. Because liquids and phase changes are complicated, we'll analyze a Carnot engine that uses an ideal gas.

The Carnot Cycle

The **Carnot cycle** is an ideal-gas cycle that consists of the two adiabatic processes ($Q = 0$) and two isothermal processes ($\Delta E_{th} = 0$) shown in Figure 19.23. These are the two types of processes allowed in a perfectly reversible gas engine. As a Carnot cycle operates,

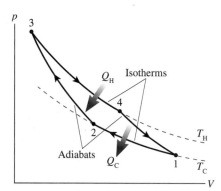

FIGURE 19.23 The Carnot cycle is perfectly reversible.

1. The gas is isothermally compressed while in thermal contact with the cold reservoir at temperature T_C. Heat energy $Q_C = |Q_{12}|$ is removed from the gas as it is compressed in order to keep the temperature constant. The compression must take place extremely slowly because there can be only an infinitesimal temperature difference between the gas and the reservoir.
2. The gas is adiabatically compressed while thermally isolated from the environment. This compression increases the gas temperature until it matches temperature T_H of the hot reservoir. No heat is transferred during this process.
3. After reaching maximum compression, the gas expands isothermally at temperature T_H. Heat $Q_H = Q_{34}$ is transferred from the hot reservoir into the gas as it expands in order to keep the temperature constant.
4. Finally, the gas expands adiabatically, with $Q = 0$, until the temperature decreases back to T_C.

Work is done in all four processes of the Carnot cycle, but heat is transferred only during the two isothermal processes.

The thermal efficiency of any heat engine is

$$\eta = \frac{W_{out}}{Q_H} = 1 - \frac{Q_C}{Q_H}$$

We can determine η_{Carnot} by finding that heat transfer in the two isothermal processes.

Process $1 \rightarrow 2$: Table 19.1 gives us the heat transfer in an isothermal process at temperature T_C:

$$Q_{12} = (W_s)_{12} = nRT_C \ln\left(\frac{V_2}{V_1}\right) = -nRT_C \ln\left(\frac{V_1}{V_2}\right) \qquad (19.22)$$

$V_1 > V_2$, so the logarithm on the right is positive. Q_{12} is negative because heat is transferred out of the system, but Q_C is simply the *amount* of heat transferred to the cold reservoir:

$$Q_C = |Q_{12}| = nRT_C \ln\left(\frac{V_1}{V_2}\right) \qquad (19.23)$$

Process 3 → 4: Similarly, the heat transferred in the isothermal expansion at temperature T_H is

$$Q_H = Q_{34} = (W_s)_{34} = nRT_H \ln\frac{V_4}{V_3} \qquad (19.24)$$

Thus the thermal efficiency of the Carnot cycle is

$$\eta_{\text{Carnot}} = 1 - \frac{Q_C}{Q_H} = 1 - \frac{T_C \ln(V_1/V_2)}{T_H \ln(V_4/V_3)} \qquad (19.25)$$

We can simplify this expression. During the two adiabatic processes,

$$T_C V_2^{\gamma-1} = T_H V_3^{\gamma-1} \quad \text{and} \quad T_C V_1^{\gamma-1} = T_H V_4^{\gamma-1} \qquad (19.26)$$

An algebraic rearrangement gives

$$V_2 = V_3\left(\frac{T_H}{T_C}\right)^{1/(\gamma-1)} \quad \text{and} \quad V_1 = V_4\left(\frac{T_H}{T_C}\right)^{1/(\gamma-1)} \qquad (19.27)$$

from which it follows that

$$\frac{V_1}{V_2} = \frac{V_4}{V_3} \qquad (19.28)$$

Consequently, the two logarithms in Equation 19.25 cancel and we're left with the result that the thermal efficiency of a Carnot engine operating between a hot reservoir at temperature T_H and a cold reservoir at temperature T_C is

$$\eta_{\text{Carnot}} = 1 - \frac{T_C}{T_H} \qquad \text{(Carnot thermal efficiency)} \qquad (19.29)$$

This remarkably simple result, an efficiency that depends only on the ratio of the temperatures of the hot and cold reservoirs, is Carnot's legacy to thermodynamics.

NOTE ▶ Temperatures T_H and T_C are *absolute* temperatures. They *must* be in kelvins. ◀

EXAMPLE 19.4 A Carnot engine
A Carnot engine is cooled by water at $T_C = 10°C$. What temperature must be maintained in the hot reservoir of the engine to have a thermal efficiency of 70%?

MODEL The efficiency of a Carnot engine depends only on the temperatures of the hot and cold reservoirs.

SOLVE The thermal efficiency $\eta_{\text{Carnot}} = 1 - T_C/T_H$ can be rearranged to give

$$T_H = \frac{T_C}{1 - \eta_{\text{Carnot}}} = 943 \text{ K} = 670°C$$

ASSESS A "real" engine would need a higher temperature than this to provide 70% efficiency because no real engine will match the Carnot efficiency.

EXAMPLE 19.5 A real engine
The heat engine of Example 19.2 had a highest temperature of 2700 K, a lowest temperature of 300 K, and a thermal efficiency of 15%. What is the efficiency of a Carnot engine operating between these two temperatures?

SOLVE The Carnot efficiency is

$$\eta_{\text{Carnot}} = 1 - \frac{T_C}{T_H} = 1 - \frac{300 \text{ K}}{2700 \text{ K}} = 0.89 = 89\%$$

ASSESS The thermodynamic cycle used in the example doesn't come anywhere close to the Carnot efficiency.

The Maximum Efficiency

In Section 19.2 we tried to invent a perfect engine with $\eta = 1$ and $Q_C = 0$. We found that we could not do so without violating the second law, so no engine can have $\eta = 1$. However, that example didn't rule out an engine with $\eta = 0.9999$. Further analysis has now shown that no heat engine operating between energy reservoirs at temperatures T_H and T_C can be more efficient than a perfectly reversible engine operating between these temperatures.

We've now reached the endpoint of this line of reasoning by establishing an exact result for the thermal efficiency of a perfectly reversible engine, the Carnot engine. We can summarize our conclusions:

> **Second law, informal statement #7** No heat engine operating between energy reservoirs at temperatures T_H and T_C can exceed the Carnot efficiency $\eta_{Carnot} = 1 - T_C/T_H$.

As Example 19.5 showed, real engines usually fall well short of the Carnot limit.

We also found that no refrigerator can exceed the coefficient of performance of a perfectly reversible refrigerator. We'll leave the proof as a homework problem, but an analysis very similar to that above shows that the coefficient of performance of a Carnot refrigerator is

$$K_{Carnot} = \frac{T_C}{T_H - T_C} \qquad \text{(Carnot coefficient of performance)} \qquad (19.30)$$

Thus we can state:

> **Second law, informal statement #8** No refrigerator operating between energy reservoirs at temperatures T_H and T_C can exceed the Carnot coefficient of performance $K_{Carnot} = T_C/(T_H - T_C)$.

EXAMPLE 19.6 Brayton versus Carnot

The Brayton-cycle refrigerator of Example 19.3 had a coefficient of performance $K = 1.1$. How does this compare to the limit set by the second law of thermodynamics?

SOLVE Example 19.3 found that the reservoir temperatures had to be $T_C \geq 250$ K and $T_H \leq 381$ K. A Carnot refrigerator operating between 250 K and 381 K has

$$K_{Carnot} = \frac{T_C}{T_H - T_C} = \frac{250 \text{ K}}{381 \text{ K} - 250 \text{ K}} = 1.9$$

ASSESS This is the minimum value of K_{Carnot}. It will be even higher if $T_C > 250$ K or $T_H < 381$ K. The coefficient of performance of the reasonably realistic refrigerator of Example 19.3 is less than 60% of the limiting value.

Statements #7 and #8 of the second law are a major result of this chapter, one with profound implications. The efficiency limit of a heat engine is set by the temperatures of the hot and cold reservoirs. High efficiency requires $T_C/T_H \ll 1$ and thus $T_H \gg T_C$. However, practical realities often prevent T_H from being significantly larger than T_C, in which case the engine cannot possibly have a large efficiency. This limit on the efficiency of heat engines is a consequence of the second law of thermodynamics.

EXAMPLE 19.7 Generating electricity

An electric power plant boils water to produce high-pressure steam at 400°C. The high-pressure steam spins a turbine as it expands, then the turbine spins the generator. The steam is then condensed back to water in an ocean-cooled heat exchanger at 25°C. What is the *maximum* possible efficiency with which heat energy can be converted to electrical energy?

MODEL The maximum possible efficiency is that of a Carnot engine operating between these temperatures.

SOLVE The Carnot efficiency depends on absolute temperatures, so we must use $T_H = 400°C = 673$ K and $T_C = 25°C = 298$ K. Then

$$\eta_{max} = 1 - \frac{298}{673} = 0.56 = 56\%$$

ASSESS This is an upper limit. Real coal-, oil-, gas-, and nuclear-heated steam generators actually operate at ≈35% thermal efficiency. (The heat *source* has nothing to do with the efficiency. All it does is boil water.) Thus, as the photo at the beginning to this chapter noted, electric power plants convert only about one-third of the fuel energy to electric energy while exhausting about two-thirds of the energy to the environment as waste heat. Not much can be done to alter the low-temperature limit. The high-temperature limit is determined by the maximum temperature and pressure the boiler and turbine can withstand. The efficiency of electricity generation is far less than most people imagine, but it is an unavoidable consequence of the second law of thermodynamics.

A limit on the efficiency of heat engines was not expected. We are used to thinking in terms of energy conservation, so it comes as no surprise that we cannot make an engine with $\eta > 1$. But the limits arising from the second law were not anticipated, nor are they obvious. Nonetheless, they are a very real fact of life and a very real constraint on any practical device. No one has ever invented a machine that exceeds the second-law limits, and we have seen that the maximum efficiency for realistic engines is surprisingly low.

STOP TO THINK 19.5   Could this heat engine be built?

a. Yes.
b. No.
c. It's impossible to tell without knowing what kind of cycle it uses.

SUMMARY

The goal of Chapter 19 has been to investigate the physical principles that govern the operation of heat engines and refrigerators.

GENERAL PRINCIPLES

Heat Engines

Devices which transform heat into work. They require two energy reservoirs at different temperatures.

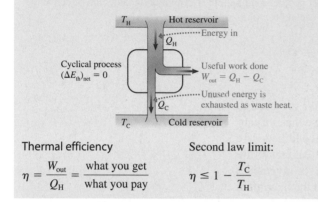

Thermal efficiency

$$\eta = \frac{W_{out}}{Q_H} = \frac{\text{what you get}}{\text{what you pay}}$$

Second law limit:

$$\eta \le 1 - \frac{T_C}{T_H}$$

Refrigerators

Devices which use work to transfer heat from a colder object to a hotter object.

Coefficient of performance

$$K = \frac{Q_C}{W_{in}} = \frac{\text{what you get}}{\text{what you pay}}$$

Second law limit:

$$K \le \frac{T_C}{T_H - T_C}$$

IMPORTANT CONCEPTS

A perfectly reversible engine (a **Carnot engine**) can be operated as either a heat engine or a refrigerator between the same two energy reservoirs by reversing the cycle and with no other changes.

- A **Carnot heat engine** has the maximum possible thermal efficiency of any heat engine operating between T_H and T_C.

$$\eta_{Carnot} = 1 - \frac{T_C}{T_H}$$

- A **Carnot refrigerator** has the maximum possible coefficient of performance of any refrigerator operating between T_H and T_C.

$$K_{Carnot} = \frac{T_C}{T_H - T_C}$$

The Carnot cycle for a gas engine consists of two isothermal processes and two adiabatic processes.

An energy reservoir is a part of the environment so large in comparison to the system that its temperature doesn't change as the system extracts heat energy from or exhausts heat energy to the reservoir. All heat engines and refrigerators operate between two energy reservoirs at different temperatures T_H and T_C.

The **work** W_s done *by* the system has the opposite sign to the work done *on* the system.

W_s = area under pV curve

APPLICATIONS

To analyze a heat engine or refrigerator:

MODEL Identify each process in the cycle.

VISUALIZE Draw the pV diagram of the cycle.

SOLVE There are several steps:

- Determine p, V, and T at the beginning and end of each process.
- Calculate ΔE_{th}, W_s, and Q for each process.
- Determine W_{in} or W_{out}, Q_H, and Q_C.
- Calculate $\eta = W_{out}/Q_H$ or $K = Q_C/W_{in}$.

ASSESS Verify $(\Delta E_{th})_{net} = 0$. Check signs.

TERMS AND NOTATION

thermodynamics	closed-cycle device	refrigerator	pressure ratio, r_p
energy reservoir	thermal efficiency, η	coefficient of performance, K	perfectly reversible engine
energy-transfer diagram	waste heat	heat exchanger	Carnot engine
heat engine			Carnot cycle

EXERCISES AND PROBLEMS

Exercises

Section 19.1 Turning Heat into Work

Section 19.2 Heat Engines and Refrigerators

1. What are (a) W_{out} and Q_C and (b) the thermal efficiency for the heat engine shown in Figure Ex19.1?

FIGURE EX19.1

2. What are (a) W_{out} and Q_H and (b) the thermal efficiency for the heat engine shown in Figure Ex19.2?

FIGURE EX19.2

3. A heat engine extracts 55 kJ of heat from the hot reservoir each cycle and exhausts 40 kJ of heat. What are (a) the thermal efficiency and (b) the work done per cycle?

4. A heat engine does 20 J of work per cycle while exhausting 30 J of waste heat. What is the engine's thermal efficiency?

5. A heat engine with a thermal efficiency of 40% does 100 J of work per cycle. How much heat is (a) extracted from the hot reservoir and (b) exhausted to the cold reservoir per cycle?

6. A refrigerator requires 20 J of work and exhausts 50 J of heat per cycle. What is the refrigerator's coefficient of performance?

7. 50 J of work are done per cycle on a refrigerator with a coefficient of performance of 4.0. How much heat is (a) extracted from the cold reservoir and (b) exhausted to the hot reservoir per cycle?

8. The power output of a car engine running at 2400 rpm is 500 kW. How much (a) work is done and (b) heat is exhausted per cycle if the engine's thermal efficiency is 20%? Give your answers in kJ.

9. A 32%-efficient electric power plant produces 900 MW of electric power and discharges waste heat into 20°C ocean water. Suppose the waste heat could be used to heat homes during the winter instead of being discharged into the ocean. A typical American house requires an average 20 kW for heating. How many homes could be heated with the waste heat of this one power plant?

10. 1.0 L of 20°C water is placed in a refrigerator. The refrigerator's motor must supply an extra 8.0 W power to chill the water to 5°C in 1.0 hr. What is the refrigerator's coefficient of performance?

Section 19.3 Ideal-Gas Heat Engines

Section 19.4 Ideal-Gas Refrigerators

11. The heat-engine cycle of Figure Ex19.11 consists of four processes. Make a chart with rows labeled A to D and columns labeled ΔE_{th}, W_s, and Q. Fill each box in the chart with +, −, or 0 to indicate whether the quantity increases, decreases, or stays the same during that process.

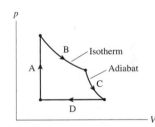

FIGURE EX19.11

12. The refrigerator cycle of Figure Ex19.12 consists of four processes. Make a chart with rows labeled A–D and columns labeled ΔE_{th}, W_s, and Q. Fill each box in the chart with +, −, or 0 to indicate whether the quantity increases, decreases, or stays the same during that process.

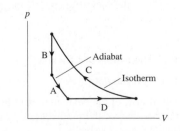

FIGURE EX19.12

13. How much work is done per cycle by a gas following the pV trajectory of Figure Ex19.13?

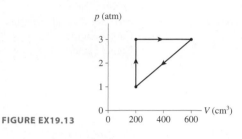

FIGURE EX19.13

14. A gas following the pV trajectory of Figure Ex19.14 does 60 J of work per cycle. What is p_{max}?

FIGURE EX19.14

15. What are (a) the thermal efficiency and (b) the heat exhausted to the cold reservoir for the heat engine shown in Figure Ex19.15?

FIGURE EX19.15

16. What are (a) the thermal efficiency and (b) the heat extracted from the hot reservoir for the heat engine shown in Figure Ex19.16?

FIGURE EX19.16

17. At what pressure ratio would a heat engine operating with a Brayton cycle have an efficiency of 60%? Assume that the gas is diatomic.

18. A heat engine uses a diatomic gas in a Brayton cycle. What is the engine's thermal efficiency if the gas volume is halved during the compression?

Section 19.5 The Limits of Efficiency

Section 19.6 The Carnot Cycle

19. Which, if any, of the heat engines in Figure Ex19.19 violate (a) the first law of thermodynamics or (b) the second law of thermodynamics? Explain.

FIGURE EX19.19

20. Which, if any, of the refrigerators in Figure Ex19.20 violate (a) the first law of thermodynamics or (b) the second law of thermodynamics? Explain.

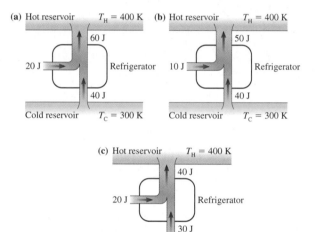

FIGURE EX19.20

21. At what cold-reservoir temperature (in °C) would a Carnot engine with a hot-reservoir temperature of 427°C have an efficiency of 60%?

22. a. A heat engine does 200 J of work per cycle while exhausting 600 J of heat to the cold reservoir. What is the engine's thermal efficiency?
    b. A Carnot engine with a hot-reservoir temperature of 400°C has the same thermal efficiency. What is the cold-reservoir temperature in °C?

23. A heat engine does 10 J of work and exhausts 15 J of waste heat during each cycle.
    a. What is the engine's thermal efficiency?
    b. If the cold-reservoir temperature is 20°C, what is the minimum possible temperature in °C of the hot reservoir?

24. A Carnot engine operating between energy reservoirs at temperatures 300 K and 500 K produces a power output of 1000 W. What are (a) the thermal efficiency of this engine, (b) the rate of heat input, in W, and (c) the rate of heat output, in W?

25. A Carnot engine whose cold-reservoir temperature is 7°C has a thermal efficiency of 40%. By how much should the temperature of the hot reservoir be increased to raise the efficiency to 60%?

26. A heat engine operating between energy reservoirs at 20°C and 600°C has 30% of the maximum possible efficiency. How much energy must this engine extract from the hot reservoir to do 1000 J of work?

27. A Carnot refrigerator operating between −20°C and +20°C extracts heat from the cold reservoir at the rate 200 J/s. What are (a) the coefficient of performance of this refrigerator, (b) the rate at which work is done on the refrigerator, and (c) the rate at which heat is exhausted to the hot side?

28. Liquid helium is kept in a refrigerator that removes heat energy from the helium at its boiling temperature of 4.2 K and exhausts heat energy into the 20°C laboratory. How much work must a Carnot refrigerator do to remove 1.0 J of heat energy from the liquid helium?

29. The coefficient of performance of a refrigerator is 5.0. The compressor uses 10 J of energy per cycle.
    a. How much heat energy is exhausted per cycle?
    b. If the hot-reservoir temperature is 27°C, what is the lowest possible temperature in °C of the cold reservoir?

30. A Carnot refrigerator with a cold-reservoir temperature of −13°C has a coefficient of performance of 5.0. To increase the coefficient of performance to 10.0, should the hot-reservoir temperature be increased or decreased, and by how much? Explain.

Problems

31. A 20%-efficient car engine accelerates the 1500 kg car from rest to 15 m/s. How much heat energy is transferred to the engine by burning gasoline?

32. The engine that powers a crane burns fuel at a flame temperature of 2000°C. It is cooled by 20°C air. The crane lifts a 2000 kg steel girder 30 m upward. How much heat energy is transferred to the engine by burning fuel if the engine is 40% as efficient as a Carnot engine?

33. 100 mL of water at 15°C is placed in the freezer compartment of a refrigerator with a coefficient of performance of 4.0. How much heat energy is exhausted into the room as the water is changed to ice at −15°C?

34. Prove that the work done in an adiabatic process i → f is $W_s = (p_f V_f − p_i V_i)/(1 − \gamma)$.

35. Statement #6 of the second law says that no refrigerator operating between reservoirs at temperatures T_H and T_C can have a coefficient of performance larger than that of a perfectly reversible refrigerator operating between these temperatures. Prove that this is true.

    Hint: The proof is a logical proof, similar to the proof of statement #5 of the second law, not a mathematical proof.

36. Prove that the coefficient of performance of a Carnot refrigerator is $K_{Carnot} = T_C/(T_H − T_C)$.

37. A Carnot heat engine with thermal efficiency η and a Carnot refrigerator operate between the same two energy reservoirs. Show that the refrigerator's coefficient of performance is $K = (1 − \eta)/\eta$.

38. You have a hot reservoir at temperature T_H, a cold reservoir at T_C, and a third energy reservoir with a temperature T_3 between T_C and T_H. A Carnot engine operating between T_H and T_C extracts heat Q_H from the hot reservoir and exhausts heat Q_C to the cold reservoir. Suppose you use two Carnot engines, one operating between T_H and T_3 and the other between T_3 and T_C. To make an appropriate comparison, the first must extract heat Q_H from the hot reservoir and the second must exhaust heat Q_C to the cold reservoir. Is the combined work done by the two engines more, less, or equal to the work of the single engine?

39. There has long been an interest in using the vast quantities of thermal energy in the oceans to run heat engines. A heat engine needs a temperature *difference,* a hot side and a cold side. Conveniently, the ocean surface waters are warmer than the deep ocean waters. Suppose you build a floating power plant in the tropics where the surface water temperature is ≈30°C. This would be the hot reservoir of the engine. For the cold reservoir, water would be pumped up from the ocean bottom where it is always ≈5°C. What is the maximum possible efficiency of such a power plant?

40. The ideal gas in a Carnot engine extracts 1000 J of heat energy during the isothermal expansion at 300°C. How much heat energy is exhausted during the isothermal compression at 50°C?

41. The hot-reservoir temperature of a Carnot engine with 25% efficiency is 80°C higher than the cold-reservoir temperature. What are the reservoir temperatures, in °C?

42. Which would make the larger increase in the thermal efficiency of a Carnot engine, raising the hot-reservoir temperature by ΔT or lowering the cold-reservoir temperature by ΔT? Explain.

43. The heat exhausted to the cold reservoir of a Carnot engine is two-thirds the heat extracted from the hot reservoir. What is the temperature ratio T_C/T_H?

44. Figure P19.44 shows a Carnot heat engine driving a Carnot refrigerator.
    a. Determine Q_1, Q_2, Q_3, and Q_4.
    b. Is Q_3 greater than, less than, or equal to Q_1?
    c. Do these two devices, when operated together in this way, violate the second law?

FIGURE P19.44

45. A heat engine running backward is called a refrigerator if its purpose is to extract heat from a cold reservoir. The same engine running backward is called a *heat pump* if its purpose is to exhaust warm air into the hot reservoir. Heat pumps are widely used for home heating. You can think of a heat pump as a refrigerator that is cooling the already cold outdoors and, with its exhaust heat Q_H, warming the indoors. Perhaps this seems a little silly, but consider the following. Electricity can be directly used to heat a home by passing an electric current through a heating coil. This is a direct, 100% conversion of work to heat. That is, 15 kW of electric power (generated by doing work at the rate 15 kJ/s at the power plant) produces

heat energy inside the home at a rate of 15 kJ/s. Suppose that the neighbor's home has a heat pump with a coefficient of performance of 5.0, a realistic value.

a. How much electric power (in kW) does the heat pump use to deliver 15 kJ/s of heat energy to the house?

b. An average price for electricity is about 40 MJ per dollar. A furnace or heat pump will run typically 200 hours per month during the winter. What does one month's heating cost in the home with a 15 kW electric heater and in the home of the neighbor who uses an equivalent heat pump?

46. Three engineering students submit their solutions to a design problem in which they were asked to design an engine that operates between temperatures 300 K and 500 K. The heat input/output and work done by their designs are shown in the following table:

Student	Q_H	Q_C	W_{out}
1	250 J	140 J	110 J
2	250 J	170 J	90 J
3	250 J	160 J	90 J

Critique their designs. Are they acceptable or not? Is one better than the others? Explain.

47. A typical coal-fired power plant burns 300 metric tons of coal *every hour* to generate 750 MW of electricity. 1 metric ton = 1000 kg. The density of coal is 1500 kg/m³ and its heat of combustion is 28 MJ/kg. Assume that *all* heat is transferred from the fuel to the boiler and that *all* the work done in spinning the turbine is transformed into electrical energy.

a. Suppose the coal is piled up in a 10 m × 10 m room. How tall must the pile be to operate the plant for one day?

b. What is the power plant's thermal efficiency?

48. A nuclear power plant generates 2000 MW of heat energy from nuclear reactions in the reactor's core. This energy is used to boil water and produce high-pressure steam at 300°C. The steam spins a turbine, which produces 700 MW of electric power, then the steam is condensed and the water is cooled to 30°C before starting the cycle again.

a. What is the maximum possible thermal efficiency of the power plant?

b. What is the plant's actual efficiency?

c. Cooling water from a river flows through the condenser (the low-temperature heat exchanger) at the rate of 1.2 × 10⁸ L/hr (≈30 million gallons per hour). If the river water enters the condenser at 18°C, what is its exit temperature?

49. The electric output of a power plant is 750 MW. Cooling water flows through the power plant at the rate 1.0 × 10⁸ L/hr. The cooling water enters the plant at 16°C and exits at 27°C. What is the power plant's thermal efficiency?

50. a. A large nuclear power plant has a power output of 1000 MW. In other words, it generates electric energy at the rate 1000 MJ/s. How much energy does this power plant supply in one day?

b. The oceans are vast. How much energy could be extracted from 1 km³ of water if its temperature is decreased by 1°C? For simplicity, assume fresh water.

c. A friend of yours who is an inventor comes to you with an idea. He has done the calculations that you just did in parts a and b, and he's concluded that a few cubic kilometers of

ocean water could meet most of the energy needs of the United States. This is an insignificant fraction of the U.S. coastal waters. In addition, the oceans are constantly being reheated by the sun, so energy obtained from the ocean is essentially solar energy. He has sketched out some design plans—highly secret, of course, because they're not patented—and now he needs some investors to provide money for a prototype. A working prototype will lead to a patent. As an initial investor, you'll receive a fraction of all future royalties. Time is of the essence because a rival inventor is working on the same idea. He needs $10,000 from you right away. You could make millions if it works out. Will you invest? If so, explain why. If not, why not? Either way, your explanation should be based on scientific principles. Sketches and diagrams are a reasonable part of an explanation.

51. Design a heat engine using a three-step closed cycle that meets the following specifications:

$$p_{min} = 1 \text{ atm} \qquad p_{max} = 3 \text{ atm}$$
$$V_{min} = 100 \text{ cm}^3 \qquad W_{out} = 40.5 \text{ J}$$

Give your answer as a pV diagram showing the cycle.

52. A heat engine using 1.0 mol of a monatomic gas follows the cycle shown in Figure P19.52. 3750 J of heat energy is transferred to the gas during process $1 \rightarrow 2$.

a. Determine W_s, Q, and ΔE for each of the four processes in this cycle. Display your results in a table.

b. What is the thermal efficiency of this heat engine?

FIGURE P19.52

53. A heat engine using a diatomic gas follows the cycle shown in Figure P19.53. Its temperature at point 1 is 20°C.

a. Determine W_s, Q, and ΔE for each of the three processes in this cycle. Display your results in a table.

b. What is the thermal efficiency of this heat engine?

c. What is the power output of the engine if it runs at 500 rpm?

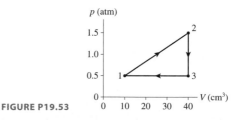

FIGURE P19.53

54. Figure P19.54 shows the cycle for a heat engine that uses a gas having $\gamma = 1.25$. The initial temperature is $T_1 = 300$ K, and this engine operates at 20 cycles per second.

a. What is the power output of the engine?

b. What is the engine's thermal efficiency?

FIGURE P19.54

55. A heat engine using a monatomic gas follows the cycle shown in Figure P19.55.
    a. Find W_s, Q, and ΔE for each process in the cycle. Display your results in a table.
    b. What is the thermal efficiency of this heat engine?

FIGURE P19.55

56. A heat engine uses a diatomic gas that follows the pV cycle in Figure P19.56.
    a. Determine the pressure, volume, and temperature at point 2.
    b. Determine ΔE_{th}, W_s, and Q for each of the three processes. Put your results in a table for easy reading.
    c. How much work does this engine do per cycle and what is its thermal efficiency?

FIGURE P19.56

57. A heat engine uses a diatomic gas that follows the pV cycle in Figure P19.57.
    a. Determine the pressure, volume, and temperature at point 1.
    b. Determine ΔE_{th}, W_s, and Q for each of the three processes. Put your results in a table for easy reading.
    c. How much work does this engine do per cycle and what is its thermal efficiency?

FIGURE P19.57

58. A Brayton-cycle heat engine follows the cycle shown in Figure P19.58. The heat input from the burning fuel is 2.0 MJ per cycle. Determine the engine's thermal efficiency by explicitly computing the work done per cycle. Compare your answer with the efficiency that you can determine from Equation 19.21.

FIGURE P19.58

59. A heat engine using 120 mg of helium as the working substance follows the cycle shown in Figure P19.59.
    a. Determine the pressure, temperature, and volume of the gas at points 1, 2, and 3.
    b. What is the engine's thermal efficiency?
    c. What is the maximum possible efficiency of a heat engine that operates between T_{max} and T_{min}?

FIGURE P19.59

60. The heat engine shown in Figure P19.60 uses 2.0 mol of a monatomic gas as the working substance.
    a. Determine T_1, T_2, and T_3.
    b. Make a table that shows ΔE_{th}, W_s, and Q for each of the three processes.
    c. What is the engine's thermal efficiency?

FIGURE P19.60

61. The heat engine shown in Figure P19.61 uses 0.020 mol of a diatomic gas as the working substance.
    a. Determine T_1, T_2, and T_3.
    b. Make a table that shows ΔE_{th}, W_s, and Q for each of the three processes.
    c. What is the engine's thermal efficiency?

FIGURE P19.61

62. A heat engine using 2.0 g of helium gas is initially at STP. The gas goes through the following closed cycle:
    - Isothermal compression until the volume is halved.
    - Isobaric expansion until the volume is restored to its initial value.
    - Isochoric cooling until the pressure is restored to its initial value.

    How much work does this engine do per cycle and what is its thermal efficiency?

63. A heat engine with 0.20 mol of a monatomic ideal gas initially fills a 2000 cm³ cylinder at 600 K. The gas goes through the following closed cycle:
    - Isothermal expansion to 4000 cm³.
    - Isochoric cooling to 300 K.
    - Isothermal compression to 2000 cm³.
    - Isochoric heating to 600 K.

    How much work does this engine do per cycle and what is its thermal efficiency?

64. Figure P19.64 is the pV diagram of Example 19.2, but now the device is operated in reverse.
    a. During which processes is heat transferred into the gas?
    b. Is this Q_H, heat extracted from a hot reservoir, or Q_C, heat extracted from a cold reservoir? Explain.
    c. Determine the values of Q_H and Q_C.

    Hint: The calculations have been done in Example 19.2 and do not need to be repeated. Instead, you need to determine which processes now contribute to Q_H and which to Q_C.
    d. Is the area inside the curve W_{in} or W_{out}? What is its value?
    e. Show that Figure 19.19 is the energy-transfer diagram of this device.
    f. The device is now being operated in a ccw cycle. Is it a refrigerator? Explain.

FIGURE P19.64

In Problems 65 through 68 you are given the equation(s) used to solve a problem. For each of these, you are to
  a. Write a realistic problem for which this is the correct equation(s).
  b. Finish the solution of the problem.

65. $0.80 = 1 - (0°C + 273)/(T_H + 273)$

66. $4.0 = Q_C/W_{in}$
    $Q_H = 100$ J

67. $0.20 = 1 - Q_C/Q_H$
    $W_{out} = Q_H - Q_C = 20$ J

68. 400 kJ $= \frac{1}{2}(p_{max} - 100$ kPa$)(3.0$ m³ $- 1.0$ m³$)$

Challenge Problems

69. Figure CP19.69 shows a heat engine going through one cycle. The gas is diatomic. The masses are such that when the pin is removed, in steps 3 and 6, the piston does not move.
    a. Draw the pV diagram for this heat engine.
    b. How much work is done per cycle?
    c. What is this engine's thermal efficiency?

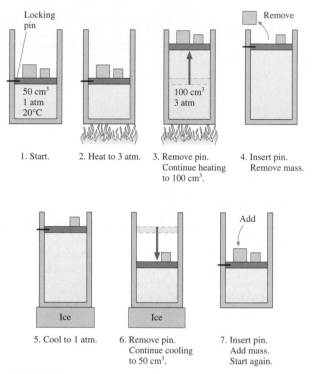

FIGURE CP19.69

70. Figure CP19.70 shows two insulated compartments separated by a thin wall. The left side contains 0.060 mol of helium at an initial temperature of 600 K and the right side contains 0.030 mol of helium at an initial temperature of 300 K. The compartment on the right is attached to a vertical cylinder, above which the air pressure is 1.0 atm. A 10-cm-diameter, 2.0 kg piston can slide without friction up and down the cylinder. Neither the cylinder diameter nor the volumes of the compartments are known.
    a. What is the final temperature?
    b. How much heat is transferred from the left side to the right side?
    c. How high is the piston lifted due to this heat transfer?
    d. What fraction of the heat is converted into work?

FIGURE CP19.70

71. The gasoline engine in your car can be modeled as the Otto cycle shown in Figure CP19.71. A fuel-air mixture is sprayed into the cylinder at point 1, where the piston is at its farthest distance from the spark plug. This mixture is compressed as the piston moves toward the spark plug during the *compression stroke*. The spark plug fires at point 2, releasing heat energy that had been stored in the gasoline. The fuel burns so quickly that the piston doesn't have time to move, so the heating is an isochoric process. The hot, high-pressure gas then pushes the piston outward during the *power stroke*. Finally, an exhaust value opens to allow the gas temperature and pressure to drop back to their initial values before starting the cycle over again.

a. Analyze the Otto cycle and show that the work done per cycle is

$$W_{out} = \frac{nR}{1 - \gamma}(T_2 - T_1 + T_4 - T_3)$$

b. Use the adiabatic connection between T_1 and T_2 and also between T_3 and T_4 to show that the thermal efficiency of the Otto cycle is

$$\eta = 1 - \frac{1}{r^{(\gamma - 1)}}$$

where $r = V_{max}/V_{min}$ is the engine's *compression ratio*.

c. Graph η versus r out to $r = 30$ for a diatomic gas.

FIGURE CP19.71

72. Figure CP19.72 shows the diesel cycle. It is similar to the Otto cycle (see Problem CP19.71), but there are two important differences. First, the fuel is not admitted until the air is fully compressed at point 2. Because of the high temperature at the end of an adiabatic compression, the fuel begins to burn spontaneously. (There are no spark plugs in a diesel engine!) Second, combustion takes place more slowly, with fuel continuing to be injected. This makes the ignition stage a constant-pressure process. The cycle shown, for one cylinder of a diesel engine, has a *displacement* $V_{max} - V_{min}$ of 1000 cm³ and a compression ratio $r - V_{max}/V_{min} = 21$. These are typical values for a diesel truck. The engine operates with intake air ($\gamma = 1.40$) at 25°C and 1.0 atm pressure. The quantity of fuel injected into the cylinder has a heat of combustion of 1000 J.

a. Find p, V, and T at each of the four corners of the cycle. Display your results in a table.

b. What is the net work done by the cylinder during one full cycle?

c. What is the thermal efficiency of this engine?

d. What is the power output in kW and horsepower (1 hp = 746 W) of an eight-cylinder diesel engine running at 2400 rpm?

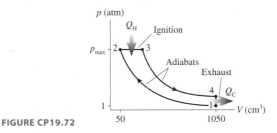

FIGURE CP19.72

STOP TO THINK ANSWERS

Stop to Think 19.1: $W_d > W_a = W_b > W_c$. $W_{out} = Q_H - Q_C$.

Stop to Think 19.2: b. Energy conservation requires $Q_H = Q_C + W_{in}$. The refrigerator will exhaust more heat out the back than it removes from the front. A refrigerator with an open door will heat the room rather than cool it.

Stop to Think 19.3: c. W_{out} = area inside triangle = 1000 J. $\eta = W_{out}/Q_H = 1000/4000 = 0.25$.

Stop to Think 19.4: To conserve energy, the heat Q_H exhausted to the hot reservoir needs to be $Q_H = Q_C + W_{out} = 40\,J + 10\,J = 50\,J$. The numbers shown here, with $Q_C = Q_H + W_{out}$, would be appropriate to a heat engine, not a refrigerator.

Stop to Think 19.5: b. The efficiency of this engine would be $\eta = W_{out}/Q_H = 0.6$. That exceeds the Carnot efficiency $\eta_{Carnot} = 1 - T_C/T_H = 0.5$, so it is not possible.

Thermodynamics

Part IV had two important goals: first, to learn how energy is transformed; second, to establish a micro/macro connection in which we can understand the macroscopic properties of solids, liquids, and gases in terms of the microscopic motions of atoms and molecules. We have been quite successful. You have learned that:

- Temperature is a measure of the thermal energy of the molecules in a system, and the average energy per molecule is simply $\frac{1}{2}k_B T$ per degree of freedom.
- The pressure of a gas is due to collisions of the molecules with the walls of the container.
- Heat is the energy transferred between two systems that have different temperatures. The *mechanism* of heat transfer is molecular collisions at the boundary between the two systems.
- Work, heat, and thermal energy can be transformed into each other in accord with the first law of thermodynamics, $\Delta E_{th} = W + Q$. This is a statement that energy is conserved.

- Practical devices for turning heat into work, called heat engines, are limited in their efficiency by the second law of thermodynamics.

The knowledge structure of thermodynamics below summarizes the basic laws, diagramming our energy model and presenting our model of a heat engine in pictorial form. Thermodynamics, more than most topics in physics, can seem very "equation oriented." It's undeniable that there are more equations than we used in earlier parts of this text and more things to remember. But focusing on the equations is seeing only the trees, not the forest. A better strategy is to focus on the ideas embedded in the knowledge structure. You can find the necessary equations if you know how the ideas are connected, but memorizing all the equations won't help if you don't know which are relevant to different situations.

KNOWLEDGE STRUCTURE IV **Thermodynamics**

ESSENTIAL CONCEPTS	Work, heat, and thermal energy.
BASIC GOALS	How is energy converted from one form to another?
	How are macroscopic properties related to microscopic behavior?

GENERAL PRINCIPLES	**First law of thermodynamics**	Energy is conserved, $\Delta E_{th} = W + Q$.
	Second law of thermodynamics	Heat is not spontaneously transferred from a colder object to a hotter object.

GAS LAWS AND PROCESSES   Ideal-gas law $pV = nRT = Nk_B T$

- Isochoric process V = constant and $W = 0$
- Isothermal process T = constant and $\Delta E_{th} = 0$

- Isobaric process p = constant
- Adiabatic process $Q = 0$

Energy Transformation

Heat Engines

W_{out} = area inside pV curve
$\quad = Q_H - Q_C$

$\eta = \dfrac{W_{out}}{Q_H}$

$\eta_{max} = \eta_{Carnot} = 1 - \dfrac{T_C}{T_H}$

Work

Requires volume change

Gas: $W = -\displaystyle\int p\,dV$

$\quad = -(\text{area under } pV \text{ curve})$

Thermal Energy

$E_{th} = \frac{1}{2}Nk_B T$ per degree of freedom

Heat

Requires temperature difference

$Q = Mc\Delta T$ or $nC\Delta T$

$Q = \pm ML$ for phase changes

Order Out of Chaos

The second law predicts that systems will run down, that order will evolve toward disorder and randomness, and that complexity will give way to simplicity. But just look around you!

- Plants grow from simple seeds to complex entities.
- Single-cell fertilized eggs grow into complex adult organisms.
- Electric current passing through a "soup" of simple random molecules produces such complex chemicals as amino acids.
- Over the last billion or so years life has evolved from simple unicellular organisms to very complex forms.
- Knowledge and information seem to grow every year, not to fade away.

Everywhere we look, it seems, the second law is being violated. How can this be?

There is an important qualification in the second law of thermodynamics: It applies only to *isolated* systems, systems that do not exchange energy with their environment. The situation is entirely different if energy is transferred into or out of the system, and we cannot predict what will happen to the entropy of a nonisolated system. The popular-science literature is full of arguments and predictions that make incorrect use of the second law by trying to apply it to systems that are not isolated.

Systems that become *more* ordered as time passes, and in which the entropy decreases, are called *self-organizing systems*. All the examples listed above are self-organizing systems. One of the major characteristics of self-organizing systems is a substantial flow of energy *through* the system. For example, plants and animals take in energy from the sun or chemical energy from food, make use of that energy, and then give waste heat back to the environment via evaporation, decay, and other means. It is this energy flow that allows the systems to maintain, or even increase, a high degree of order and a very low entropy.

But—and this is the important point—the entropy of the *entire* system, including the earth and the sun, undergoes a significant *increase* so as to let selected subsystems decrease their entropy and become more ordered. The second law is not violated at all, but you must apply the second law to the combined systems that are interacting and not just to a single subsystem.

The snowflake in the photo is a beautiful example. As water freezes, the random motion of water molecules is transformed into a highly ordered crystal. The entropy of the water molecules certainly decreases, but water doesn't

A snowflake is a highly ordered arrangement of water molecules. The creation of a snowflake decreases the entropy of the water, but the second law of thermodynamics is not violated because the water molecules are not an isolated system.

freeze as an isolated system. For it to freeze, heat energy must be transferred from the water to the surrounding air. The entropy of the air increases by *more* than the entropy of the water decreases. Thus the *total* entropy of the water + air system increases when a snowflake is formed, just as the second law predicts.

Self-organization is closely related to nonlinear mechanics, chaos, and the geometry of fractals. It has important applications in fields ranging from ecology to computer science to aeronautical engineering. For example, the air flow across a wing gives rise to large-scale turbulence—eddies and whirlpools—in the wake behind an airplane. Their formation affects the aerodynamics of the plane and can also create hazards for following aircraft. Whirlpools are ordered, large-scale macroscopic structures with low entropy, but they are produced from disordered, random collisions of the air molecules.

Self-organizing systems are a very active field of research in both science and engineering. The 1977 Nobel Prize in chemistry was awarded to the Belgian scientist Ilya Prigogine for his studies of *nonequilibrium thermodynamics*, the basic science underlying self-organizing systems. Prigogine and others have shown how energy flow through a system can, when the conditions are right, "bring order out of chaos."

Mathematics Review

Algebra

Using exponents:

$$a^{-x} = \frac{1}{a^x} \qquad a^x a^y = a^{(x+y)} \qquad \frac{a^x}{a^y} = a^{(x-y)} \qquad (a^x)^y = a^{xy}$$

$$a^0 = 1 \qquad a^1 = a \qquad a^{1/n} = \sqrt[n]{a}$$

Fractions:

$$\left(\frac{a}{b}\right)\left(\frac{c}{d}\right) = \frac{ac}{bd} \qquad \frac{a/b}{c/d} = \frac{ad}{bc} \qquad \frac{1}{1/a} = a$$

Logarithms:

If $a = e^x$, then $\ln(a) = x$ $\qquad \ln(e^x) = x \qquad\qquad e^{\ln(x)} = x$

$$\ln(ab) = \ln(a) + \ln(b) \qquad \ln\left(\frac{a}{b}\right) = \ln(a) - \ln(b) \qquad \ln(a^n) = n\ln(a)$$

The expression $\ln(a + b)$ cannot be simplified.

Linear equations: The graph of the equation $y = ax + b$ is a straight line. a is the slope of the graph. b is the y-intercept.

Proportionality: To say that y is proportional to x, written $y \propto x$, means that $y = ax$, where a is a constant. Proportionality is a special case of linearity. A graph of a proportional relationship is a straight line that passes through the origin. If $y \propto x$, then

$$\frac{y_1}{y_2} = \frac{x_1}{x_2}$$

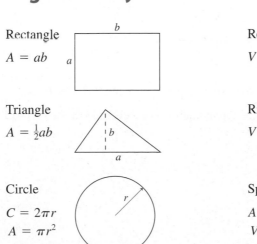

Slope $a = \dfrac{\text{rise}}{\text{run}} = \dfrac{\Delta y}{\Delta x}$

y-intercept $= b$

Quadratic equation: The quadratic equation $ax^2 + bx + c = 0$ has the two solutions $x = \dfrac{-b \pm \sqrt{b^2 - 4ac}}{2a}$.

Geometry and Trigonometry

Area and volume:

Rectangle

$A = ab$

Triangle

$A = \frac{1}{2}ab$

Circle

$C = 2\pi r$

$A = \pi r^2$

Rectangular box

$V = abc$

Right circular cylinder

$V = \pi r^2 l$

Sphere

$A = 4\pi r^2$

$V = \frac{4}{3}\pi r^3$

Arc length and angle: The angle θ in radians is defined as $\theta = s/r$.

The arc length that spans angle θ is $s = r\theta$.

$2\pi \text{ rad} = 360°$

Right triangle: Pythagorean theorem $c = \sqrt{a^2 + b^2}$ or $a^2 + b^2 = c^2$

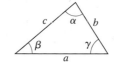

$$\sin\theta = \frac{b}{c} = \frac{\text{far side}}{\text{hypotenuse}} \qquad \theta = \sin^{-1}\left(\frac{b}{c}\right)$$

$$\cos\theta = \frac{a}{c} = \frac{\text{adjacent side}}{\text{hypotenuse}} \qquad \theta = \cos^{-1}\left(\frac{a}{c}\right)$$

$$\tan\theta = \frac{b}{a} = \frac{\text{far side}}{\text{adjacent side}} \qquad \theta = \tan^{-1}\left(\frac{b}{a}\right)$$

General triangle: $\alpha + \beta + \gamma = 180° = \pi \text{ rad}$

Law of cosines $c^2 = a^2 + b^2 - 2ab\cos\gamma$

Identities:

$$\tan\alpha = \frac{\sin\alpha}{\cos\alpha} \qquad\qquad \sin^2\alpha + \cos^2\alpha = 1$$

$$\sin(-\alpha) = -\sin\alpha \qquad\qquad \cos(-\alpha) = \cos\alpha$$

$$\sin(\alpha \pm \beta) = \sin\alpha\cos\beta \pm \cos\alpha\sin\beta \qquad \cos(\alpha \pm \beta) = \cos\alpha\cos\beta \mp \sin\alpha\sin\beta$$

$$\sin(2\alpha) = 2\sin\alpha\cos\alpha \qquad\qquad \cos(2\alpha) = \cos^2\alpha - \sin^2\alpha$$

$$\sin(\alpha \pm \pi/2) = \pm\cos\alpha \qquad\qquad \cos(\alpha \pm \pi/2) = \mp\sin\alpha$$

$$\sin(\alpha \pm \pi) = -\sin\alpha \qquad\qquad \cos(\alpha \pm \pi) = -\cos\alpha$$

Expansions and Approximations

Binomial expansion: $(1 + x)^n = 1 + nx + \dfrac{n(n-1)}{2}x^2 + \ldots$

Binomial approximation: $(1 + x)^n \approx 1 + nx \quad \text{if} \quad x \ll 1$

Trigonometric expansions: $\sin\alpha = \alpha - \dfrac{\alpha^3}{3!} + \dfrac{\alpha^5}{5!} - \dfrac{\alpha^7}{7!} + \ldots \text{ for } \alpha \text{ in rad}$

$$\cos\alpha = 1 - \frac{\alpha^2}{2!} + \frac{\alpha^4}{4!} - \frac{\alpha^6}{6!} + \ldots \text{ for } \alpha \text{ in rad}$$

Small-angle approximation: If $\alpha \ll 1$ rad, then $\sin\alpha \approx \tan\alpha \approx \alpha$ and $\cos\alpha \approx 1$.

The small-angle approximation is excellent for $\alpha < 5°$ (≈ 0.1 rad) and generally acceptable up to $\alpha \approx 10°$.

Periodic Table of Elements

Period

Group																	
1 1 H 1.0																	2 He 4.0
2 3 Li 6.9	4 Be 9.0											5 B 10.8	6 C 12.0	7 N 14.0	8 O 16.0	9 F 19.0	10 Ne 20.2
3 11 Na 23.0	12 Mg 24.3											13 Al 27.0	14 Si 28.1	15 P 31.0	16 S 32.1	17 Cl 35.5	18 Ar 39.9
4 19 K 39.1	20 Ca 40.1	21 Sc 45.0	22 Ti 47.9	23 V 50.9	24 Cr 52.0	25 Mn 54.9	26 Fe 55.8	27 Co 58.9	28 Ni 58.7	29 Cu 63.5	30 Zn 65.4	31 Ga 69.7	32 Ge 72.6	33 As 74.9	34 Se 79.0	35 Br 79.9	36 Kr 83.8
5 37 Rb 85.5	38 Sr 87.6	39 Y 88.9	40 Zr 91.2	41 Nb 92.9	42 Mo 95.9	43 Tc 96.9	44 Ru 101.1	45 Rh 102.9	46 Pd 106.4	47 Ag 107.9	48 Cd 112.4	49 In 114.8	50 Sn 118.7	51 Sb 121.8	52 Te 127.6	53 I 126.9	54 Xe 131.3
6 55 Cs 132.9	56 Ba 137.3	57 La 133.9	72 Hf 178.5	73 Ta 180.9	74 W 183.9	75 Re 186.2	76 Os 190.2	77 Ir 192.2	78 Pt 195.1	79 Au 197.0	80 Hg 200.6	81 Tl 204.4	82 Pb 207.2	83 Bi 209.0	84 Po 209.0	85 At 210.0	86 Rn 222.0
7 87 Fr 223.0	88 Ra 226.0	89 Ac 227.0	104 Rf 261	105 Db 262	106 Sg 263	107 Bh 264	108 Hs 269	109 Mt 268	110 Ds 271	111 272	112 285						

— Transition elements —

Atomic number — 27
Co — Symbol
Atomic mass — 58.9

Lanthanides 6

58 Ce 140.1	59 Pr 140.9	60 Nd 144.2	61 Pm 144.9	62 Sm 150.4	63 Eu 152.0	64 Gd 157.3	65 Tb 158.9	66 Dy 162.5	67 Ho 164.9	68 Er 167.3	69 Tm 168.9	70 Yb 173.0	71 Lu 175.0

Actinides 7

90 Th 232.0	91 Pa 231.0	92 U 238.0	93 Np 237.0	94 Pu 239.1	95 Am 241.1	96 Cm 244.1	97 Bk 249.1	98 Cf 252.1	99 Es 257.1	100 Fm 257.1	101 Md 258.1	102 No 259.1	103 Lr 262.1

— Inner transition elements —

Answers

Answers to Odd-Numbered Exercises and Problems

Solutions to questions posed in the Part Overview captions can be found at the end of this answer list.

Chapter 16

1. 22.6 m^3
3. 8.33 cm
5. 4.82×10^{23} atoms
7. a. 6.02×10^{28} atoms/m^3 b. 3.28×10^{28} atoms/m^3
9. 2.17 cm
11. Lowest: $-88°C = 185 \text{ K}$; highest $58°C = 331 \text{ K}$
13. a. $171°Z$ b. $671°C = 944 \text{ K}$
15. a. $32.02°F$, 608 Pa b. $-68.8°F$, $5.06 \times 10^5 \text{ Pa}$
17. Freezing point lower, boiling point higher.
19. a. 0.0497 m^3 b. 1.33 atm
21. 18.8 atm
23. a. 55.1 mol b. 1.234 m^3
25. a. 5.41×10^{23} atoms b. 3.59 g c. $2.30 \times 10^{26} \text{ m}^{-3}$ d. 1.52 kg/m^3
27. a. 0.732 atm b. 0.523 atm
29. a. 9520 kPa
    b.
31. a. 12.02 atm
    b.
33. a. Isothermal b. Both are 914 K c. 300 cm^3
35. 0.228 nm
37. a. $7.03 \times 10^{-21} \text{ J}$ b. 2060 m/s
39. $1.1 \times 10^{15} \text{ m}^{-3}$
41. a. 1.32×10^{-13} b. 1.24×10^{11} molecules
43. $174°C$
45. 92.8 cm^3
47. 34.7 psi
49. $174.3°C$
51. a. 3.05×10^{21} b. 2.02 mg
53. No

55.
57. a. 884 kPa b. $323°C$, $-49.5°C$, $397.5°C$
59. a. Both $366 \text{ K} = 93°C$ b. Isothermal c. $1098 \text{ K} = 825°C$
61. a. 0.509 atm b. $-112°C$
    c.
63. a. 4.0 atm, $-73°C$
    b.

65. b.

   c. 6 atm
67. b.

   c. 417 cm³
69. a. 23.5 cm b. 7.8 cm
71. a. 2.73 m b. 10.96 atm
73. 1.02 cm

Chapter 17

1. 490 J
3. 40 J
5. 200 cm³
7.

9.

11.

13. 700 J from the system
15. 12,500 J
17. 6864 J
19. 6.79×10^4 J
21. 27.5°C
23. 73.5°C
25. Iron
27. a. 91.2 J b. 140°C
29. a. 1.14 atm b. 48.5°C
31. a. 26.4 b. 7.07
33. 8.73 hr
35. 994 cm³
37. −56.4°C
39. Aluminum
41. 87.3 min
43. a. 83.3 J/kg K   b. 2.0×10^5 J/kg
45. 0.0605 kg = 60.5 g
47. 5450 J
49. a. 245 J b. 944 m/s
51. 1660 J
53. a. 253°C b. 32.6 cm
55. 7750 J
57. a. 4290 cm³, 606°C b. 3050 J c. 1.0 atm d. 2180 J
   e.

59. A: −1010 J, B: 1419 J
61. a. 811 J b. −487 J c. 0 J
63. a. A: 2.46×10^{-3} m³, 300 K; B: 1.80×10^{-3} m³, 220 K
   b.

65. a. −50.7 J b. −24.8 J c. 25.9 J
67. 1100 K, 23.9 cm³
69. a. 39.3 b. 171
71. a. 0.5 atm b. −1074 J c. 1074 J d. 0 J
   e.

73. a. 5460 K b. 0 J c. 5.39×10^4 J d. 20
 e.

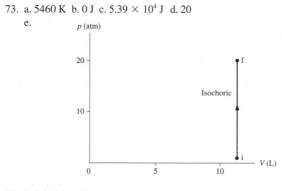

75. b. 0.0156 mol
77. b. 0.472 kg
79. a. At 1: 3.0 atm, 946°C, 1000 cm^3. At 2: 1.0 atm, 946°C, 3000 cm^3.
 At 3: 0.48 atm, 310°C, 3000 cm^3. b. -334 J, 0 J, 239 J
 c. 334 J, -239 J, 0 J
81. 14.5 atm

Chapter 18

1. 2.69×10^{25} m^{-3}
3. a. 3.30×10^{12} m^{-3} b. 1.71×10^6 m
5. 61
7. a. $0\hat{\imath} + 0\hat{\jmath}$ b. 59.2 m/s c. 61.6 m/s
9. a. 9.16×10^4 Pa b. 332 K
11. 1.91×10^{24} s^{-1}
13. 2820 m/s, 891 m/s
15. 283 m/s
17. a. 68.3 K b. 1090 K
19. 7.22×10^{12} K
21. a. 3400 J b. 3400 J c. 3400 J
23. a. 4×10^{-16} J b. 7×10^5 m/s
25. 3.65×10^7 J
27. a. 0.0800°C b. 0.0481°C c. 0.0400°C
29. a. 62.4 J b. 104 J c. 104 J d. 145 J
31. a. Gas B b. A: 5200 J, B: 7800 J
33. 8.48
35. a. Helium b. 1367 m/s c. 1.86×10^{-6} m
37. 9.6×10^{-5} m/s
39. a. $\lambda = 1/[\sqrt{2}\,\pi(N/V)r^2]$ b. 1.82×10^{-6} Pa or 1.80×10^{-11} atm
41. a. 1.273×10^{25} m^{-3} b. 449 m/s c. 259 m/s d. 1.296×10^{25} s^{-1}
 e. 56,800 Pa f. 56,700 Pa
43. a. Helium: 30.4 J; Argon: 121.6 J b. Helium: 47.3 J; Argon: 104.7 J
 c. 16.9 J is transferred from the argon to the helium. d. 580 K
 e. Helium: 3.11 atm; Argon: 3.45 atm
45. 482 K
51. a. $R = 8.31$ J/mol K b. $2R = 16.6$ J/mol K
53. a. 4 b. 1 c. 16
55. 1/2
57. a. 2.03×10^6 J b. 4.83×10^{-6} c. 0.00132 K
59. a.   b. $9p_i V_i$

61. c. 436 K, 850 J is transferred from the oxygen to the helium.

Chapter 19

1. a. 10 J, 110 J b. 0.0833
3. a. 0.273 b. 15 kJ
5. a. 250 J b. 150 J
7. a. 200 J b. 250 J
9. 96,000
11.

	ΔE_{th}	W_s	Q
A	+	0	+
B	0	+	+
C	−	+	0
D	−	−	−

13. 40.5 J
15. a. 0.0952 b. 285 J
17. 24.7
19. a. b only b. a only
21. 7°C
23. a. 40% b. 215°C
25. 233 K
27. a. 6.32 b. 32 W c. 232 J/s
29. a. 60 J b. -23°C
31. 8.44×10^5 J
33. 5.34×10^4 J
39. 8.25%
41. 47°C and -33°C
43. 2/3
45. a. 2.5 kW b. $270 and $45
47. a. 48 m b. 32.1%
49. 37%
51.

p (atm) diagram

or

p (atm) diagram

53. a.

	W_s (J)	Q (J)	ΔE_{th} (J)
1 → 2	3.04	16.97	13.93
2 → 3	0	−10.13	−10.13
3 → 1	−1.52	−5.32	−3.80
Net	1.52	1.52	0

 b. 8.96% c. 12.7 W
55. a.

	W_s (J)	Q (J)	ΔE_{th} (J)
1 → 2	0	282.2	282.2
2 → 3	207.2	0	−207.2
3 → 1	−50.0	−125.0	−75.0
Net	157.2	157.2	0

 b. 52.2%

57. a. 5.743×10^4 Pa, 4000×10^{-6} m^3, 229.7 K
  b.

	ΔE_{th} (J)	W_s (J)	Q (J)
$1 \rightarrow 2$	425.7	-425.7	0
$2 \rightarrow 3$	0	554.5	554.5
$3 \rightarrow 1$	-425.7	0	-425.7
Net	0	128.8	128.8

  c. 23.2%
59. a. Point 1: 1.013×10^5 Pa, 1.0×10^{-3} m^3, 406 K;
   Point 2: 5.06×10^5 Pa, 1.0×10^{-3} m^3, 2030 K;
   Point 3: 1.013×10^5 Pa, 5.0×10^{-3} m^3, 2030 K b. 28.8% c. 80%
61. a. 1620 K, 2407 K, 6479 K
  b.

	ΔE_{th} (J)	W_s (J)	Q (J)
$1 \rightarrow 2$	327	-327	0
$2 \rightarrow 3$	1692	677	2369
$3 \rightarrow 1$	-2019	0	-2019
Net	0	350	350

  c. 14.8%

63. 345.6 J, 24.0%
65. b. $T_H = 1092°C$
67. b. $Q_H = 100$ J, $Q_C = 80$ J
69. a.

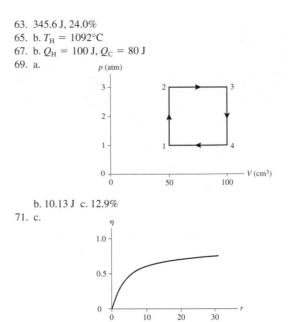

  b. 10.13 J c. 12.9%
71. c.

Part Overview Solutions

PART IV Overview

R2-D2's internal systems have been on standby while inside the ice cave on the planet Hoth. How much energy must R2-D2's fuel cells provide to raise the temperature of his internal systems from the standby temperature of $-20°C$ to a working temperature of $30°C$? To find out, what properties of R2-D2 do you need to estimate?

MODEL Assume that R2–D2 is well insulated so that all the energy goes into raising his temperature, and none of the energy escapes into the ice cave. Although R2-D2 is made of many parts, they are mostly aluminum (structural materials) and silicon (semiconductor electronics). Thus we'll estimate that the *average* specific heat is 800 J/kg K. R2-D2 isn't terribly large, so we'll estimate the mass of his internal systems to be 40 kg.

SOLVE The heat energy needed to change the temperature by ΔT is

$$Q = mc\Delta T = (40 \text{ kg})(800 \text{ J/kg K})(30°C - (-20°C))$$

$$= 1.6 \text{ MJ}$$

ASSESS The conversion of electrical energy from the fuel cells into heat energy is nearly 100% efficient. Thus the fuel cells must supply 1.6 MJ of energy.

Credits

All Part Overview images are courtesy of Lucasfilm Ltd. Addison Wesley would like to give special thanks to Lucy Wilson, Christopher Holm, and the staff of Lucasfilm Ltd. for granting us permission to use these images and for their help in selecting them.

TITLE PAGE
Page **iii**: Rainbow/PictureQuest.

PART IV
Part IV Overview image: *Star Wars: Episode V – The Empire Strikes Back* © 1980 and 1997 Lucasfilm Ltd. & ™. All rights reserved. Used under authorization. Unauthorized duplication is a violation of applicable law.

CHAPTER 16
Page **485**: Luis Veiga/Getty Images. Page **486**: Kevin Schafer/Corbis. Page **489**: Richard Megna/Fundamental Photos. Page **491** T: David Young-Wolff/PhotoEdit. Page **491** B: David Young-Wolff/PhotoEdit.

CHAPTER 17
Page **512**: W.A. Sharman/Milepost 92/Corbis. Page **514**: Robert & Linda Mostyn/Eye Ubiquitous/Corbis. Page **520**: Kevin Fleming/Corbis. Page **529**: Roger Ressmeyer/Corbis.

CHAPTER 18
Page **547**: Pete Turner/Getty Images.

CHAPTER 19
Page **573**: Mark Wagner/Getty Images. Page **574**: Spencer Grant/PhotoEdit. Page **575**: W. Cody/Corbis. Page **577**: Larry Chiger/SuperStock. **581**: Carrier Corporation. Page **586**: Malcolm Fife/Getty Images. Page **589**: Paul Silverman/Fundamental Photos.

PART IV SUMMARY
Page **607**: Kristian Hilsen/Getty Images.

Index

Q

R